Lecture Notes in Computer Science 16067

Founding Editors

Gerhard Goos
Juris Hartmanis

The series Lecture Notes in Computer Science (LNCS), including its subseries Lecture Notes in Artificial Intelligence (LNAI) and Lecture Notes in Bioinformatics (LNBI), has established itself as a medium for the publication of new developments in computer science and information technology research, teaching, and education.

LNCS enjoys close cooperation with the computer science R & D community, the series counts many renowned academics among its volume editors and paper authors, and collaborates with prestigious societies. Its mission is to serve this international community by providing an invaluable service, mainly focused on the publication of conference and workshop proceedings and postproceedings. LNCS commenced publication in 1973.

Matthijs van Leeuwen · Jilles Vreeken

Editors

Challenges and Algorithms for Knowledge Discovery from Data

Essays Dedicated to Arno Siebes
on the Occasion of His 67th Birthday

 Springer

Editors
Matthijs van Leeuwen (iD)
Leiden University
Leiden, The Netherlands

Jilles Vreeken (iD)
CISPA Helmholtz Center for Information
Security
Saarbrücken, Germany

ISSN 0302-9743   ISSN 1611-3349 (electronic)
Lecture Notes in Computer Science
ISBN 978-3-032-03027-6   ISBN 978-3-032-03028-3 (eBook)
https://doi.org/10.1007/978-3-032-03028-3

Preface

June 27th, 2025, marked the 67th birthday of Prof. Dr. Arno Siebes, full professor of Algorithmic Data Analysis at Utrecht University. In the Netherlands, reaching the age of 67 has a special meaning for full professors, as it implies a mandatory transition from full professor to professor emeritus, which is a friendly way to say that the time to retire has come. As we are confident that Arno thinks more positively about reaching the age of 67 than about retirement, this *liber amicorum* (or *Festschrift*) is specifically dedicated to the occasion of the former.

One of the many customs in Dutch academia is for full professors to give an inaugural lecture when they get their appointment, as well as to give a farewell lecture when they retire. "Patterns…what else?"[1] was the title of Arno's farewell symposium and lecture on September 22–23, 2025. This book was presented to Arno at his farewell symposium, as a surprise and present from his colleagues in the European data mining and machine learning community.

Arno's academic career started at Utrecht University in 1977, where he studied Mathematics and graduated in 1983. After a brief stint at the Tax Administration, he joined CWI in Amsterdam in 1985. He obtained his Ph.D. in 1990 from Twente University for his dissertation 'On Complex Objects', which he completed under the supervision of Peter Apers and Martin Kersten. After his Ph.D. he stayed in the Database Group of Martin Kersten at the CWI. In 1999 he became a part-time full professor at TU Eindhoven. The lure of Utrecht remained strong, however, and he returned to his Alma Mater already in 2000 to take the chair for Large Distributed Databases. This chair was renamed to Algorithmic Data Analysis in 2007. Retirement in 2025 means that Arno was a full professor at Utrecht University for exactly 25 years. Throughout his career Arno supervised 15 Ph.D. students to a successful defense. Four of those, meanwhile, hold the title of professor and have together supervised 33 Ph.D. students so far. That is, Arno has a legacy of over 48 academic descendants, and this number is expected to grow steeply, as at the time of writing, also three of his *academic grandchildren* recently became professors and others share this ambition.

Arno's research interests can be catch-phrased as 'theory that works in practice'. With a strong background in databases, he focused on the question 'What *insights* can we extract from this data?' and sought after solutions that are not just beautiful on paper but also work in the sense that they do something new that existing methods cannot. With this in mind, it is no surprise that he became a driving force in the European data mining community. One branch of research where he was particularly visible is that of *inductive databases*. These are databases where not just the data, but also the results of data mining algorithms are first-class citizens and can be queried as such. These were,

---

[1] As an aside, from the title one may get the impression that—following George Clooney's example—Arno may pursue a commercial career after having become famous as an academic, and we certainly believe his flamboyant appearance would be a valuable asset in such an endeavour.

among others, extensively studied by the subsequent EU projects KESO, CinQ, and IQ, in which Arno was closely involved.

Out of all topics, pattern mining holds the most special place in Arno's heart. The title of the symposium, "Patterns...what else?", does not come as a surprise to anyone who knows his research. He may not have been the one to propose the problem of pattern mining, but he is arguably the one who solved its most pressing problem: the pattern explosion. Arno realized that asking for *all* patterns that satisfy some local constraints is the wrong question when we actually want a succinct and non-redundant set of patterns that together generalize the data well. And, if this is what we are after, we should phrase the question as such and use an inductive principle to measure the quality of the *set of patterns* as a whole. The final ingredient was the observation that patterns describe local parts of the data, and, hence the Minimum Description Length (MDL) principle is a natural choice for identifying the best pattern set. This highly influential idea was first presented at the SIAM Conference on Data Mining in 2006, and its initial incarnation is now known as the KRIMP algorithm. The ideas of pattern set mining and using MDL for data mining have been picked up by many and have been extended to a broad range of settings and data mining problems, from itemsets to graph structures, and from classification to causal inference.

Arno has always been an influential member of the data mining community—many of the contributions in this book speak highly of him in this regard. A key example of his lasting impact is that in 2001 he was program chair of the first co-located edition of the European Conference on Machine Learning (ECML) and the European Conference on Principles and Practice of Knowledge Discovery in Databases (PKDD) in Freiburg, Germany, together with Luc De Raedt. The experiment was a great success; in the European spirit the communities behind ECML and PKDD grew ever closer together, formally merged into ECML PKDD in 2008, and is still a thriving and ever-growing community. Arno has been a highly visible member of the ECML PKDD community, both in attending the conference and community meetings, as member of the Steering Committee, and as its treasurer. Arno has also played an active and pivotal role in the Symposium on Intelligent Data Analysis (IDA), which he attended and co-chaired numerous times in the past decades, where he always advocated for ideas over performance.

As for this book, we—as editors—created a list of current and past close collaborators of Arno currently active in the field, and invited all of them to contribute a chapter. We received 13 submissions, all of which were reviewed by two authors of other submissions (i.e., each lead author was assigned two other submissions and could invite others as subreviewers). The review process was single-blind. All reviews were positive and provided detailed feedback for further improvement. Based on this advice, we accepted all submissions, and encouraged the authors to use the feedback to further improve their contributions.

This resulted in the chapters in this book, which are categorized into three topical sections. Naturally, we start with a section on pattern mining, the research field that Arno arguably has contributed most to (until now, at least). Unsurprisingly, a highly frequent pattern in this section is MDL-based pattern set mining. Then, we switch to the broader topic of learning and reasoning, which Arno has certainly also enjoyed and contributed to. Finally, we wrap up with a section on a topic that is very timely but at first glance seems surprising in the context of our honoree: large language models. The last chapter, however, will demonstrate how well large language models and Arno go together.

Dear Arno, the contributions in this book make it abundantly clear what a large and lasting impact you have had on the European data mining and machine learning community, in many regards. We hope that this book will serve as a memory and reminder of that impact, but also as an encouragement to keep in touch with the people in that community. All the best!

July 2025

Matthijs van Leeuwen
Jilles Vreeken

Organization

Program Committee Chairs

Matthijs van Leeuwen | Leiden University, the Netherlands
Jilles Vreeken | CISPA Helmholtz Center for Information Security, Germany

Program Committee

Hendrik Blockeel | KU Leuven, Belgium
Robert Castelo | Universitat Pompeu Fabra, Spain
Bruno Cremilleux | Université de Caen Normandie, France
Luc De Raedt | KU Leuven, Belgium
Ad Feelders | Utrecht University, Netherlands
Elisa Fromont | Université de Rennes 1, France
Johannes Fürnkranz | Johannes Kepler University Linz, Austria
Bart Goethals | University of Antwerp, Belgium
Andrej Kastrin | University of Ljubljana, Slovenia
Katharina Morik | TU Dortmund, Germany
Céline Robardet | INSA Lyon, France
Blaž Škrlj | Jožef Stefan Institute, Slovenia

Additional Reviewers

Wouter Duivesteijn
Sébastien Ferré
Robin Manhaeve
Giuseppe Marra
Alexandre Termier
Albrecht Zimmermann

Contents

Large Language Models

Pattern Mining

A Subjective Pattern Mining Literature Survey

Jean-François Boulicaut[1], Marc Plantevit[2], and Céline Robardet[1(✉)]

[1] INSA Lyon, CNRS, LIRIS UMR 5205, 69621 Villeurbanne, France
celine.robardet@insa-lyon.fr
[2] EPITA Research Laboratory (LRE), 94276 Le Kremlin-Bicêtre, France

Abstract. This survey paper covers about three decades of research on pattern discovery. We consider some of the major breakthroughs about constraint-based data mining and several nice results about interesting pattern set discovery. We then focus on recent approaches for a cross-fertilization between pattern discovery and neural-based Machine Learning approaches. Specifically, neural methods have been developed for pattern mining and pattern set mining, enabling more efficient exploration of structural motifs in complex datasets. Additionally, hybrid approaches integrating Explainable AI (XAI) techniques have emerged, aiming to enhance the interpretability and transparency of machine learning models. We finally illustrate to what extent our colleague Arno Siebes has contributed to the considered topics.

Keywords: Data Mining · Pattern discovery · Neural approaches for pattern mining · XAI

1 Introduction

In the 1990s, it became evident that huge amounts of collected data remained largely underutilized. This awareness spurred the emergence of the Data Mining research domain, which focuses on uncovering patterns and regularities in data [2,3]. Positioned between database querying and machine learning or statistical methods, data mining introduced a complementary approach for extracting knowledge from data, ultimately enhancing the value derived from collected information. Early data mining efforts focused on symbolic data, a type of data that had previously received relatively little attention. The idea was to harness advances in computational power and algorithmic efficiency to extract insights from large datasets, an area that had been largely underexplored by traditional statistical methods. Interesting patterns, defined by constraints, were discovered using exact and correct algorithms. Their efficiency stemmed from the use of carefully designed order relations for pattern enumeration, which enabled the effective pruning of large portions of the search space.

At that time, data mining research results have proven highly effective for processing symbolic data and extracting patterns that are easily interpretable by humans. Their ability to identify frequent itemsets, association rules, and other structured knowledge representations has made it a valuable tool for various applications, from market basket analysis to bioinformatics. At the same time, more structured data types, such as relational and graph-based data, have become increasingly abundant, particularly with the rise of social networks, biological networks, and knowledge graphs. This shift has led to the development of numerous techniques specifically designed for subgraph pattern mining and network analysis. These approaches, useful to uncover recurring structures, communities, and relationships within complex networks, have been applied in fraud detection, drug discovery, recommendation systems, and beyond.

However, in particularly challenging scenarios, such as when constraints hinder effective pruning, it became necessary to relax the requirements of completeness and correctness. In these cases, approximation strategies, including heuristic methods, were explored. This shift often entailed moving beyond the discovery of isolated local patterns to the identification of meaningful sets of patterns. As a result, new algorithmic challenges emerged, prompting two decades of active research into pattern set mining techniques. These approaches have proven especially valuable in application domains where sets of local patterns contribute to the construction of more or less global models.

More recently, the impressive advances made in neural networks and their representation learning capabilities have had a significant impact on the field of data mining. The ability of deep learning models to automatically learn meaningful representations of complex data offers opportunities and raises several important questions at the intersection of these two domains. First, can the representations learned by neural networks be leveraged for pattern mining? Traditional data mining techniques focus on extracting human-interpretable patterns from symbolic or structured data. With neural networks learning high-dimensional embeddings of data points, an intriguing possibility is to use these representations as a basis for pattern discovery [78]. Second, can pattern-based approaches help us understand neural networks? One of the key challenges in modern Artificial Intelligence is the lack of interpretability in deep learning models. Pattern mining techniques, which are designed inherently to extract meaningful structures from data, can serve as a tool to analyze and explain the decision-making processes of neural networks [73]. By identifying recurring substructures, feature interactions, and decision rules within a trained model, we can gain insights into how it processes information. Finally, can differentiable programming be used to compute pattern sets and address challenges in pattern set mining? Classical pattern mining techniques often struggle with scalability and combinatorial explosion, particularly when searching for collections of patterns rather than individual patterns. Differentiable programming, that enables optimization through gradient-based methods, can be used for pattern mining [28,32,77]. It leverages continuous optimization techniques instead of traditional combinatorial approaches.

This chapter revisits nearly three decades of data mining research by highlighting several key milestones. We begin with foundational work centered on the development of exact, correct, and complete algorithms for local pattern discovery (Sect. 2). We then explore advances in constructing global, interpretable models learned in an unsupervised manner from patterns (Sect. 3), which in some ways foreshadow the convergence between data mining and deep learning. Finally, we examine the integration of data mining with machine learning and artificial intelligence, underscoring the growing interplay and mutual reinforcement among these fields (Sect. 4). We conclude by acknowledging the significant contributions of our colleague Arno Siebes to the development and evolution of these research domains.

2 Inductive Queries to Support Local Pattern Discovery

A fundamental abstraction for many data mining tasks was introduced in [49]. Within the constraint-based data mining framework, the discovery of interesting patterns from a pattern language $\mathcal{L}$ in a dataset $\mathcal{D}$ is formalized as computing:

$$Q = \{\varphi \in \mathcal{L} \mid \mathcal{C}(\varphi, \mathcal{D}) \text{ holds}\}$$

where $\mathcal{C}(\varphi, \mathcal{D})$ is a constraint to be satisfied. Typically, $\mathcal{C}$ is constructed as a conjunction of primitive constraints relevant to the pattern domain. For instance, in frequent itemset mining [3], $\mathcal{D}$ represents transactional (binary) data, $\mathcal{L}$ is the set of itemsets, and $\mathcal{C}$ enforces a minimum frequency threshold. Constraints like $\mathcal{C}$ can specify both syntactic and semantic requirements for patterns, as well as guide their a priori interestingness.

From a computational viewpoint, a naïve approach based on enumerating all candidates from $\mathcal{L}$ and filtering them using $\mathcal{C}$, is generally intractable. Consequently, much research has focused on developing ad hoc algorithms that push constraints into the search process. For more general-purpose approaches, the main challenge becomes exploiting constraint properties to guide and optimize enumeration. This paradigm was extensively explored in the 2000s, especially through European projects on inductive databases [38,61], which yielded significant advances consolidated in works like [30,50].

2.1 Local Pattern Discovery

Local patterns include itemsets and association rules, formal concepts, subsequences and episodes, data dependencies, subtrees, and subgraphs [51]. A compelling example is the discovery of formal concepts, modeled as maximal rectangles in Boolean matrices where rows represent objects and columns represent attributes. Here, a pattern is a pair comprising a maximal object set and a maximal attribute set. To guarantee correctness, a closedness constraint ensures that each object set is associated with the maximal attribute set shared by its elements, and vice versa. Additional constraints (e.g., on the minimum number of objects or attributes) can help filter trivial patterns.

Such constraints can also incorporate fault-tolerance, allowing exceptions that account for noise or variability. More complex pattern structures, as those in n-ary relations, can be discovered by generalizing maximal rectangles [25–27]. In short, declarative constraints do not merely define the pattern space, they also encode interestingness criteria.

2.2 Computation and Constraint Properties

The challenge of computing Q exactly and completely is most evident in frequent itemset mining, where constraints include a minimal frequency threshold. The classic Apriori algorithm [3] addresses this by employing a level-wise search strategy and exploiting the downward closure property: if an itemset is infrequent, all its supersets are also infrequent.

To enable efficient pruning, constraint properties like monotonicity and anti-monotonicity are key. A constraint is monotone if any superset of a satisfying pattern also satisfies it. It is anti-monotone if any superset of a non-satisfying pattern also fails the constraint. These properties support early termination of search branches. Moreover, their conjunctions preserve monotonicity or anti-monotonicity, allowing for bidirectional pruning when mixed.

Other constraint classes include: Succinct constraints, which can be evaluated before mining begins; Convertible constraints [57], whose monotonicity depends on the enumeration order; Loose anti-monotone constraints [15], where a pattern of size k satisfying the constraint implies at least one $(k-1)$-subset does as well, ensuring that all necessary information is available to verify the constraint.

While monotonic and anti-monotonic constraints can be effectively exploited to prune the search space using both depth-first and breadth-first algorithms, loose anti-monotonic constraints generally require breadth-first exploration. However, this limitation can be overcome by using reverse search algorithms [4,36,66], which enable depth-first traversal even in the presence of loose anti-monotonicity. These algorithms define a reduction function that assigns a unique parent to each pattern, ensuring a complete and memory-efficient exploration of the search space. In general, depth-first strategies are advantageous in terms of memory usage, making them particularly suitable for large or complex pattern spaces.

When constraints are non-monotone, algorithms can still leverage boundable constraints [52], which admit either monotone upper or lower bounds. While effective, the computation of such bounds is often ad hoc, meaning that a specific bound must be manually derived for each individual problem. To achieve greater generality, piecewise monotone and anti-monotone constraints [26,27] have been proposed. These approaches decompose complex constraints into simpler components that are locally monotone, thereby enabling effective pruning even in the absence of global monotonicity. Other works have also explored the idea of decomposing complex constraints into simpler locally monotone sub-constraints to facilitate efficient search and pruning [22,63].

2.3 Condensed Representation

A significant challenge in pattern mining is the phenomenon of pattern flooding, where the number of extracted patterns becomes too large to interpret or use effectively. One strategy to address this issue is through the use of condensed representations, particularly with respect to frequency. The idea is to compute a subset $\mathcal{CR} \subseteq \mathcal{L}$ that is as compact as possible, while still allowing for the efficient reconstruction of the full result set Q. In the context of large-scale data mining, "efficient" means that this reconstruction should be possible without any additional access to the original dataset.

A well-known example of such an approach involves the use of border sets [49], such as the set of maximal frequent itemsets. These itemsets are considered a good condensed representation because every subset of a maximal frequent itemset is itself frequent. Therefore, the full set of frequent itemsets can be inferred from the maximal ones without scanning the data again. This makes the maximal frequent itemsets a proper and efficient subset of Q.

However, in most real-world applications, users are interested not only in identifying which patterns are frequent but also in knowing their support values, as these are essential for computing measures of interestingness, such as the confidence of association rules. Consequently, a condensed representation must allow not only for the regeneration of the patterns themselves but also for the accurate retrieval of evaluation functions like support. If this regeneration is exact, the representation is referred to as an exact condensed representation; otherwise, if the values are only approximated, it is known as an approximate condensed representation.

The usefulness of a condensed representation is generally evaluated based on several criteria. These include the size of the representation, both in theoretical terms and in practice, the computational efficiency and completeness of the algorithms that generate it, and the ease and speed with which useful information, such as the set of all frequent itemsets and their support values, can be recovered from it. Over the years, many useful condensed representations have been developed. These include closed itemsets, δ-free itemsets, disjunction-free sets, non-derivable itemsets, and k-free itemsets, with a detailed survey available in [24]. A more recent overview of the concept is provided in [64].

In addition to their summarization role, closed sets are widely recognized for their ability to enhance the efficiency of pattern extraction. Their strong pruning properties drastically reduce the size of the search space, making large-scale mining tasks not only feasible but also more manageable. This is especially relevant when mining frequent patterns or user-centric subjective patterns in large or complex datasets.

3 Pattern Set Mining

An early finding in data mining research was that the mining process typically generates extensive collections of patterns, making them ineffective for the end-user. To address this limitation, a huge effort was made to reduce the size of the output by focusing on actually interesting patterns with constraint-based pattern mining, condensed representations, and statistical assessments of the patterns.

Another important direction in the field has been to shift the focus from evaluating patterns in isolation to assessing the usefulness of collections of patterns. Rather than considering each pattern independently, these approaches aim to extract sets of patterns that, taken together, provide more informative and interpretable insights for the end-user.

Individual patterns have often been viewed as "fragmented knowledge," lacking a coherent structure or clear relationships to one another. This limitation has sparked growing interest in pattern set mining, the task of identifying collections of patterns that jointly satisfy certain desirable properties [59]. To tackle challenges such as redundancy and overwhelming output size, various strategies have emerged, including pattern teams [40], constraint-based pattern set mining [59], and pattern selection techniques [17]. These approaches aim to improve the interpretability and practical utility of mined results by promoting diversity, minimizing redundancy, and emphasizing relevance.

3.1   Local Patterns to Global Models

A first significant attempt to compute a collection of patterns that collectively optimize a global objective is tiling [35]. The goal of tiling is to identify the smallest set of tiles (i.e., itemsets) that effectively cover the 1 s (positive entries) in a binary dataset. A good solution must balance two key objectives: minimizing the overlap between tiles while ensuring high-quality coverage of the dataset. This problem has been extended to ranked data [67], where the challenge lies in adapting tiling methods to accommodate order constraints and ranking dependencies inherent in the data. In [54], tiling is formulated as an Integer Linear Programming (ILP) problem, enabling the use of ILP solvers to compute optimized solutions.

Another important perspective in pattern set mining is the construction of global models based on local patterns, as formalized in the LEGO framework (Local Patterns to Global Models) [33,39]. The LEGO approach consists of three main phases: Local Pattern Discovery, Pattern Set Discovery, and Global Modeling. In the first phase, the search space is explored using user-defined inductive constraints to generate candidate patterns. Each pattern is then evaluated individually using quality measures such as frequency or predictive performance for a target concept. The second phase selects a compact, informative, and non-redundant subset of patterns from this potentially large set. Finally, in the Global Modeling phase, the selected patterns are transformed into a coherent model, either by treating them as constructed features and applying standard inductive methods, or through a pattern-specific strategy tailored to the pattern type.

3.2   Pattern Set Mining Based on Minimum Description Length Principle

An elegant way to build a relevant pattern set is to focus on extracting patterns that efficiently compress the dataset [62, 70, 76]. The core idea is that the patterns that compress the data well are typically the most informative. By leveraging the Minimum Description Length (MDL) principle, KRIMP algorithm identifies a small and non-redundant set of patterns that effectively summarize the data.

KRIMP optimizes the Minimum Description Length which states that the best model for a dataset is the one that minimizes the total description length, which includes both the cost of encoding the model (pattern set) and the cost of encoding the data using that model. The KRIMP algorithm seeks to optimize the formula:

$$L(D, C) = L(C) + L(D \mid C)$$

where $L(D, C)$ represents the total description length, $L(C)$ corresponds to the length required to encode the code table C, which consists of a set of patterns, and $L(D \mid C)$ denotes the length required to encode the data D using the code table C. The objective is to find a code table C that minimizes $L(D, C)$, balancing the complexity of the code table with its effectiveness in encoding the dataset.

The first component, $L(C)$, measures the cost of encoding the code table itself. The code table serves as a model for the dataset and consists of patterns, which are itemsets that frequently appear together in the data. Each pattern is assigned a code length, determined by its usage frequency, with shorter codes assigned to more frequent patterns. A smaller and more concise code table reduces this encoding cost while maintaining its ability to effectively compress the data.

The second component, $L(D \mid C)$, represents the cost of encoding the dataset D using the patterns in the code table. Each transaction in D is encoded as a combination of patterns from C, typically following a greedy strategy. This means that encoding starts with the most specific patterns, those that cover the largest portions of a transaction, before using more general patterns until the entire transaction is covered. The goal is to represent D as efficiently as possible, reducing the total number of bits required for encoding.

The objective of KRIMP is thus to find the code table C that minimizes the total description length $L(D, C)$. This ensures that: (i) The code table C is concise since it avoids including unnecessary or redundant patterns. (ii) The data D is encoded efficiently since patterns are chosen to provide the best compression of the dataset. The MDL principle assumes that the patterns that compress the data best are likely to capture the most meaningful and relevant structure in the dataset. By balancing $L(C)$ (model complexity) and $L(D|C)$ (data fit), KRIMP avoids overfitting or including overly complex models.

This pioneering work has paved many research directions and MDL stands as an effective and versatile technique. MDL has been successfully applied to handdle missing values [74], to capture the norm and the differences between several datasets [18, 75], to discover correlations [19], or causalities [20, 21]. It was also successfully used for detecting changes in streams [69] and to build interpretable rule-based classifier [58]. MDL principle has also been applied to other types of patterns. In [23], Calders et al. apply it to process mining. The MDL principle has been extensively applied to sequential data, to discover sequential patterns [12, 42, 43], periodic sequences [34], sequential rules [16], patterns within time series [72], and within event sequences [11]. Finally, some researchers has also investigated the MDL principle to mine pattern in graphs [6, 7] and knowledge graphs [8].

3.3   Subjective Interestingness for Pattern Set Discovery

Another approach to deriving non-redundant and informative pattern sets is based on the subjective interestingness framework introduced by Tijl De Bie [13]. This framework leverages Maximum Entropy models and Information Theory to evaluate how surprising or informative a pattern is with respect to the user's prior knowledge and expectations.

At the heart of this approach lies the construction of a maximum entropy model, which captures the user's prior beliefs about the data. This model represents the least biased probability distribution that satisfies known constraints, such as marginal distributions or frequencies of certain patterns. It serves as a baseline against which observed patterns are evaluated. A pattern is deemed interesting if it significantly deviates from the model's predictions, quantified through information gain, reflecting the difference between the observed and expected probabilities of a pattern. Importantly, the framework does not simply select the pattern with the highest information content. Instead, it incorporates the assimilation cost of a pattern, typically measured by its description length. The most interesting pattern is then the one that best balances information content against cognitive or representational cost.

To compute a pattern set, the process is iterative: once the most informative pattern is identified and presented to the user, the user's prior is updated through the maximum entropy model. This updating process progressively refines the model's understanding of what is surprising, thus enabling the discovery of increasingly relevant patterns aligned with the user's evolving expectations.

The subjective interestingness framework has been successfully applied well beyond the itemset mining domain. Applications include real-valued data [41], exploratory data analysis [14], multi-relational data [46, 65], subgroup discovery with continuous targets [45], trees [1], antichains in hierarchies [9], subgraphs [68], and attributed graphs [10].

4   Bridging Pattern Mining and Machine Learning

These longstanding data mining challenges are increasingly being revisited through the lens of neural approaches. In particular, neural methods have been developed for both pattern mining and pattern set mining, enabling more efficient discovery of structural motifs in complex datasets. Moreover, hybrid approaches that integrate techniques from Explainable AI (XAI) have emerged, with the goal of enhancing the interpretability and transparency of machine learning models.

4.1 Representation Learning for Pattern Mining

Deep learning approaches have demonstrated impressive capabilities in learning rich representations from complex data, which can be effectively leveraged in downstream tasks. In the context of graph data, recent advances have explored the use of deep learning techniques to tackle combinatorially hard problems in graph analysis, such as edit distance prediction, graph isomorphism testing, maximum common subgraph identification, and frequent subgraph mining. Traditional combinatorial algorithms for these tasks, such as A* search for edit distance computation or the Weisfeiler-Lehman test for graph isomorphism, often exhibit high computational complexity, rendering them impractical for large-scale graphs. Neural approaches seek to overcome these limitations by employing Graph Neural Networks (GNNs) and embedding-based representations to approximate solutions with greater efficiency. For instance, [5] proposed a neural method for graph edit distance prediction, which learns to estimate the minimum number of edit operations required to transform one graph into another. This is achieved by embedding graphs into a latent space where distances between embeddings correlate with their edit distances. Similarly, several works [31,37,44] have focused on graph isomorphism testing, designing GNN architectures that learn expressive representations capable of distinguishing non-isomorphic graphs. Many of these methods extend the Weisfeiler-Lehman test to improve their discriminative power. Another key challenge is the pairwise maximum common subgraph (MCS) problem, which aims to find the largest subgraph common to two input graphs. Meanwhile, substructure counting methods [29,47] focus on estimating the number of occurrences of specific motifs or subgraphs within a given graph. These techniques are particularly relevant for applications in network science, cheminformatics, and bioinformatics, where identifying recurrent substructures is crucial. Most of these methods rely on neural message-passing frameworks to efficiently capture and propagate structural patterns.

This line of research, that consists in applying neural approaches to combinatorial problems, paves the way for computing patterns under constraints using neural methods. For instance, [78] proposed leveraging representation learning for frequent subgraph mining through a novel approach based on order-embedding spaces. Order embeddings are a representation learning technique designed to capture partial order structures via geometric relationships among embeddings [71]. In this framework, the preserved order relation in the embedding space corresponds to the subgraph ordering, such that the position of a graph's embedding encodes its hierarchical relationship to other subgraphs.

To achieve this, the method employs a Graph Neural Network (GNN) to learn embeddings that encode structural information while respecting the subgraph ordering constraints. Once the embeddings are precomputed, the approach enables efficient subgraph property determination: at query time, it becomes possible to verify whether one graph is a subgraph of another in linear time with respect to the embedding dimension. This embedding-based framework is further integrated with an enumeration procedure that estimates the frequency of subgraphs based on their learned representations. By leveraging the geometric properties of the embedding space, the method can efficiently retrieve the most frequent subgraphs while significantly reducing the computational complexity compared to traditional combinatorial subgraph mining techniques. This approach opens new perspectives for scalable and data-driven subgraph mining in large networks.

4.2   Neural Approaches for Pattern Set Mining

Pattern set mining has traditionally aimed at identifying meaningful patterns within datasets, focusing on extracting subsets that are both representative and non-redundant. However, the computational complexity of such pattern sets increases super-exponentially with dataset size, making exact enumeration methods impractical for large datasets. As previously discussed, approximate approaches have been explored to overcome these limitations. More recently, neural methods have emerged as an alternative, leveraging the computational power of modern machine learning frameworks, GPU acceleration, and differentiable optimization techniques to improve scalability and efficiency.

Fischer and Vreeken [32] introduce a novel approach that integrates pattern set mining into a neural framework by making the process differentiable. Their method is based on an autoencoder architecture with a binary latent space, where each neuron in the hidden layer represents a conjunctive pattern. This structure enables the model to learn interpretable patterns while maintaining the benefits of an end-to-end differentiable optimization process. The autoencoder consists of a single hidden pattern layer and an output layer that reconstructs the dataset based on the extracted patterns. The encoder and decoder share weights, ensuring consistency in how patterns are represented and reconstructed.

To guarantee that the hidden neurons correspond to meaningful patterns, the model imposes additional constraints. A discrete negative bias is applied to the pattern layer, enforcing a minimum activation threshold so that patterns only activate when a sufficient number of relevant features are present in the input. Additionally, a weight mirroring constraint ensures that the weight matrix of the decoding layer is simply the transposed version of the encoding layer, preserving a fixed input-output relationship. This prevents arbitrary activations and guarantees that each pattern neuron reconstructs only the subset of data points it represents.

The training process relies on a hybrid approach where the forward pass operates with binarized weights and activations, while the backward pass updates continuous versions of the weights to allow for smooth gradient-based optimization. This technique enables the model to learn efficiently while maintaining discrete representations in inference. Moreover, the approach dynamically determines the optimal number of patterns by initializing with a large number of hidden neurons and allowing the network to gradually prune unnecessary ones. Connections to redundant neurons naturally decay to zero-weighted edges, leading to an adaptive and compact pattern representation.

Diffnaps [77] and DiffVersify [28] extend the differentiable binary autoencoder approach to the task of subgroup discovery, where the goal is to identify patterns that succinctly describe and distinguish between different classes in the data. Diffnaps achieves this by integrating a binary autoencoder with a classification module attached to the hidden layer, enabling joint optimization of both reconstruction and classification objectives. The classification component consists of a fully connected layer directly linked to the pattern layer of the autoencoder, ensuring that the learned latent representations capture symbolic patterns that are both interpretable and discriminative. To encourage the discovery of class-specific patterns, Diffnaps optimizes a multi-task loss function. The autoencoder is trained to minimize reconstruction error, while the classification head is optimized to reduce misclassification. This joint training process naturally steers the network towards learning patterns that are not only representative of the data structure but also crucial for distinguishing between classes. By enforcing a logistic regression-based classification on the pattern layer, Diffnaps ensures that the presence or absence of specific patterns becomes predictive of class membership, enhancing the model's interpretability and effectiveness in subgroup discovery.

While Diffnaps successfully extracts class-discriminative patterns, it faces challenges when the number of underlying classes or features increases, potentially leading to redundancy in the learned patterns and leaving parts of the data uncovered. DiffVersify addresses these limitations by introducing a novel regularization mechanism that promotes both better data coverage and pattern diversity. The key innovation lies in an orthogonality-based regularization term incorporated into the loss function, which encourages the extracted patterns to be more diverse and complementary, reducing redundancy and improving the model's robustness in complex datasets. Additionally, DiffVersify enhances the decoding process by replacing the conventional thresholding techniques used in previous methods with a more flexible and expressive non-negative matrix factorization (NMF)-based decoding strategy. This approach refines the way patterns are reconstructed, allowing for a more nuanced interpretation of the learned representations. The combination of improved pattern diversity and an advanced decoding mechanism significantly enhances the model's performance, making DiffVersify more adaptable to a wide range of datasets while maintaining interpretability and scalability.

A major advantage of these neural formulations lies in their scalability. While traditional pattern mining methods often exhibit exponential complexity with respect to the number of attributes, deep learning models typically rely on polynomial-time algorithms for both training and inference, in terms of the number of parameters and training examples—provided that exact convergence to the global minimum of the loss function is not required. Moreover, deep learning computations can be significantly accelerated by leveraging the massive parallelism of GPUs.

By framing pattern set mining as a differentiable optimization problem, these approaches bridge the gap between symbolic and connectionist paradigms, offering a powerful alternative for discovering structured knowledge in large-scale datasets, while still preserving a degree of interpretability.

4.3   Pattern Mining and XAI

Neural approaches are not only used to compute constrained patterns or pattern sets, but can also be leveraged to extract symbolic knowledge or to construct high-accuracy self-interpretable models.

For instance, [55] introduces LACE, a model-agnostic method for structured data that identifies influential features and feature values affecting a classifier's predictions. It evaluates feature importance by omitting attributes and measuring the resulting change in prediction probability. Similar to Shapley values, it captures feature interactions, highlighting combinations that jointly impact predictions. Since exhaustively computing all possible subsets is infeasible, LACE relies on a local interpretable model trained on K-nearest neighbors of the instance, labeled by the original model. It then uses an associative classifier to extract relevant patterns, significantly reducing computational complexity while maintaining interpretability. Similarly, [56] investigates performance variations across different data subsets. Using subgroup discovery, the method identifies subsets with specific attribute values where model performance, measured by false positives or false negatives, significantly deviates from overall dataset metrics. By revealing patterns that impact predictive reliability, this approach helps diagnose biases and improve model fairness.

Veyrin-Forrer et al. [73] introduced INSIDE-GNN, a pattern mining approach operating in the latent space of Graph Neural Networks (GNNs). This method isolates internal features embedded within the network layers that are automatically learned by the GNN for graph classification. By analyzing hidden layers, INSIDE-GNN identifies sets of neurons that exhibit distinct activation patterns depending on the target variable. These activation rules capture specific configurations in the embedding space that are crucial for the GNN's decision-making process. Additionally, they reveal latent structural features of input graphs that contribute to model predictions. INSIDE-GNN explains GNN predictions by extracting activation rules from the network's latent space, enabling both instance-level and model-level explanations. Since the learned subspaces are not directly interpretable, the approach generates representative graphs that embody the extracted rules. To achieve this, a Monte Carlo Tree Search (MCTS)-based method constructs realistic graphs that fully adhere to the identified activation patterns. Furthermore, the authors propose an alternative strategy based on computing the median graph of the supporting instances, demonstrating that this method produces representative and higher-quality visualizations of the discovered patterns.

In a similar vein, Livanos et al. [48] propose an approach to analyzing deep neural networks by identifying what they term a concept backbone, a subgraph of hidden units that frequently co-activate for a specific subset of instances associated with a particular concept. These concepts can represent various phenomena, such as a group of misclassified instances or other patterns of interest that warrant explanation. Their method relies on pattern mining techniques to extract meaningful itemsets from neural activations, coupled with a heuristic designed to construct a diverse and non-redundant set of concept backbones. This ensures that the extracted subgraphs correspond to different concepts while remaining distinct from one another. By leveraging these structured representations, the approach provides deeper insights into how neural networks encode and process information related to different aspects of a dataset.

Pattern mining has recently been integrated into differentiable machine learning models to develop self-interpretable predictive frameworks. One notable example is LAGRA (Learning Attributed GRAphlets) [60], a model based on the extraction and combination of recurring attributed subgraphs to enhance both interpretability and predictive performance. By integrating graph mining with machine learning, LAGRA identifies and utilizes meaningful subgraph patterns for making predictions. In this approach, the structural components of the subgraphs are discovered using a graph mining algorithm that systematically explores the dataset to extract relevant subgraph patterns. Unlike traditional methods where attribute values are predefined, LAGRA treats them as continuous trainable parameters, allowing the model to learn optimal attribute representations that contribute most effectively to predictions. To ensure efficiency and interpretability, LAGRA model is linearly combination of attributed subgraphs. It uses a L1 sparsity constraint on the coefficients to promote the selection of a compact yet informative subset of subgraphs while eliminating redundant patterns. An efficient pruning mechanism is incorporated by combining proximal gradient updates, which enforce sparsity, with graph mining tree search, a systematic exploration technique for identifying the most relevant subgraphs. This combination allows LAGRA to scale effectively to large graphs while maintaining high predictive accuracy and interpretability.

5   Arno Siebes: A Mind that Mines Wisdom

Arno Siebes is not only a respected and prolific data mining researcher but also an impressive academic who has always spent time and energy to train young researchers and organize our research communities. On the scale of his whole career, one finds evidence that his solid background in mathematics and statistics but also about databases has been inspiring many original contributions to data mining and machine learning. We already cited some of his papers, starting with the proposal in 1995 of an inductive query language [61]. It was a nice step towards what has been studied later on as an inductive database perspective. He made also important contributions to pattern set mining based on the Minimum Description Length principle (see Sect. 3) which have paved the research direction followed by many researchers to combine machine learning and pattern mining (Sect. 4). Even though Arno Siebes has been involved mostly in data mining basic research, he has also been involved in important application domains (see, e.g., [53]).

His services for our research communities have been remarkable. Without claiming to be exhaustive, he has been the co-organizer of two Dagstuhlh seminars on Local Pattern Discovery and two of the six international workshop on Inductive Databases. Arno Siebes has been the first co-chair with Luc de Raedt of ECML PKDD. It was in 2001 and they both convinced the ECML community (European Conference on Machine Learning that have been launched in 1993) and the PKDD community (Principles of Knowledge Discovery in Databases, say data mining, that has been launched in 1997) that a joint event would be a plus due to the amazing cross-fertilization possibilities between both communities. Since then, the annual meeting ECML PKDD is a major venue in Europe for researchers and practitioners. Arno Siebes has been also among the promoters of the Intelligent Data Analysis symposium (23rd symposium in 2025). This is a focused meeting that emphasizes the impact of statistics, data mining and machine learning for modern Data Analysis.

The authors of this paper have invaluable memories of many meetings with Arno Siebes. Just to name some of them, one thinks of conference or workshop venues, Ph. D defenses and social events in Antwerp, Berlin, Brussels, Cavtat, Freiburg, Grenoble, Helsinki, Leiden, Leuven, London, Ljubjana, Lyon, Montpellier, Paris, Pisa, Porto, Utrecht and Wadern. We had so many occasions during the last two decades, consistently feeling smarter after than before each of these meetings: This is what happens when exposed to brilliant, though humble scholars.

References

1. Adriaens, F., Lijffijt, J., De Bie, T.: Subjectively interesting connecting trees and forests. Data Min. Knowl. Disc. **33**(4), 1088–1124 (2019). https://doi.org/10.1007/s10618-019-00627-1

2. Agrawal, R., Imielinski, T., Swami, A.N.: Mining association rules between sets of items in large databases. In: Buneman, P., Jajodia, S. (eds.) Proceedings of the 1993 ACM SIGMOD International Conference on Management of Data, Washington, DC, USA, 26–28 May 1993, pp. 207–216. ACM Press (1993)

3. Agrawal, R., Srikant, R.: Fast algorithms for mining association rules in large databases. In: J.B., Jarke, M., Zaniolo, C. (eds.) VLDB 1994, Proceedings of 20th International Conference on Very Large Data Bases, 12–15 September 1994, Santiago de Chile, Chile, pp. 487–499. Morgan Kaufmann (1994)

4. Avis, D., Fukuda, K.: Reverse search for enumeration. Discret. Appl. Math. **65**(1–3), 21–46 (1996)

5. Bai, Y., Ding, H., Bian, S., Chen, T., Sun, Y., Wang, W.: SimGNN: a neural network approach to fast graph similarity computation. In: Culpepper, J.S., Moffat, A., Bennett, P.N., Lerman, K. (eds.) Proceedings of the Twelfth ACM International Conference on Web Search and Data Mining, WSDM 2019, Melbourne, VIC, Australia, 11–15 February 2019, pp. 384–392. ACM (2019)

6. Bariatti, F., Cellier, P., Ferré, S.: GraphMDL: graph pattern selection based on minimum description length. In: Berthold, M.R., Feelders, A., Krempl, G. (eds.) IDA 2020. LNCS, vol. 12080, pp. 54–66. Springer, Cham (2020). https://doi.org/10.1007/978-3-030-44584-3_5
7. Bariatti, F., Cellier, P., Ferré, S.: GraphMDL+: interleaving the generation and mdl-based selection of graph patterns. In: Hung, C.-C., Hong, J., Bechini, A., Song, E. (eds.) SAC 2021: The 36th ACM/SIGAPP Symposium on Applied Computing, Virtual Event, Republic of Korea, 22–26 March 2021, pp. 355–363. ACM (2021)
8. Bariatti, F., Cellier, P., Ferré, S.: KG-MDL: mining graph patterns in knowledge graphs with the MDL principle. CoRR, abs/2309.12908 (2023)
9. Bendimerad, A., Lijffijt, J., Plantevit, M., Robardet, C., De Bie, T.: Contrastive antichains in hierarchies. In: Teredesai, A., Kumar, V., Li, Y., Rosales, R., Terzi, E., Karypis, G. (eds.) Proceedings of the 25th ACM SIGKDD International Conference on Knowledge Discovery & Data Mining, KDD 2019, Anchorage, AK, USA, 4–8 August 2019, pp. 294–304. ACM (2019)
10. Bendimerad, A., Mel, A., Lijffijt, J., Plantevit, M., Robardet, C., De Bie, T.: SIAS-miner: mining subjectively interesting attributed subgraphs. Data Min. Knowl. Discov. **34**(2), 355–393 (2020)
11. Bertens, R., Vreeken, J., Siebes, A.: Keeping it short and simple: summarising complex event sequences with multivariate patterns. CoRR, abs/1512.07056 (2015)
12. Bhattacharyya, A., Vreeken, J.: Efficiently summarising event sequences with rich interleaving patterns. In: Chawla, N.V., Wang, W. (eds.) Proceedings of the 2017 SIAM International Conference on Data Mining, Houston, Texas, USA, 27–29 April 2017, pp. 795–803. SIAM (2017)
13. De Bie, T.: Maximum entropy models and subjective interestingness: an application to tiles in binary databases. Data Min. Knowl. Discov. **23**(3), 407–446 (2011)
14. Bie, T.: Subjective interestingness in exploratory data mining. In: Tucker, A., Höppner, F., Siebes, A., Swift, S. (eds.) IDA 2013. LNCS, vol. 8207, pp. 19–31. Springer, Heidelberg (2013). https://doi.org/10.1007/978-3-642-41398-8_3
15. Bonchi, F., Lucchese, C.: Extending the state-of-the-art of constraint-based pattern discovery. Data Knowl. Eng. **60**(2), 377–399 (2007)
16. Bourrand, E., Galárraga, L., Galbrun, E., Fromont, É., Termier, A.: Discovering useful compact sets of sequential rules in a long sequence. In: 33rd IEEE International Conference on Tools with Artificial Intelligence, ICTAI 2021, Washington, DC, USA, 1–3 November 2021, pp. 1295–1299. IEEE (2021)
17. Bringmann, B., Zimmermann, A.: The chosen few: on identifying valuable patterns. In: Proceedings of the 7th IEEE International Conference on Data Mining (ICDM 2007), 28–31 October 2007, Omaha, Nebraska, USA, pp. 63–72. IEEE Computer Society (2007)
18. Budhathoki, K., Vreeken, J.: The difference and the norm—characterising similarities and differences between databases. In: Appice, A., Rodrigues, P.P., Santos Costa, V., Gama, J., Jorge, A., Soares, C. (eds.) ECML PKDD 2015. LNCS (LNAI), vol. 9285, pp. 206–223. Springer, Cham (2015). https://doi.org/10.1007/978-3-319-23525-7_13
19. Budhathoki, K., Vreeken, J.: Correlation by compression. In: Chawla, N.V., Wang, W. (eds.) Proceedings of the 2017 SIAM International Conference on Data Mining, Houston, Texas, USA, 27–29 April 2017, pp. 525–533. SIAM (2017)
20. Budhathoki, K., Vreeken, J.: MDL for causal inference on discrete data. In: Raghavan, V., Aluru, S., Karypis, G., Miele, L., Wu, X. (eds.) 2017 IEEE International Conference on Data Mining, ICDM 2017, New Orleans, LA, USA, 18–21 November 2017, pp. 751–756. IEEE Computer Society (2017)

21. Budhathoki, K., Vreeken, J.: Origo: causal inference by compression. Knowl. Inf. Syst. **56**(2), 285–307 (2018)
22. Buzmakov, A., Kuznetsov, S.O., Napoli, A.: Fast generation of best interval patterns for nonmonotonic constraints. In: Appice, A., Rodrigues, P.P., Santos Costa, V., Gama, J., Jorge, A., Soares, C. (eds.) ECML PKDD 2015. LNCS (LNAI), vol. 9285, pp. 157–172. Springer, Cham (2015). https://doi.org/10.1007/978-3-319-23525-7_10
23. Calders, T., Günther, C.W., Pechenizkiy, M., Rozinat, A.: Using minimum description length for process mining. In: Shin, S.Y., Ossowski, S. (eds.) Proceedings of the 2009 ACM Symposium on Applied Computing (SAC), Honolulu, Hawaii, USA, 9–12 March 2009, pp. 1451–1455. ACM (2009)
24. Calders, T., Rigotti, C., Boulicaut, J.-F.: A survey on condensed representations for frequent sets. In: Boulicaut, J.-F., De Raedt, L., Mannila, H. (eds.) Constraint-Based Mining and Inductive Databases. LNCS (LNAI), vol. 3848, pp. 64–80. Springer, Heidelberg (2006). https://doi.org/10.1007/11615576_4
25. Cerf, L., Besson, J., Nguyen, K.-N., Boulicaut, J.-F.: Closed and noise-tolerant patterns in n-ary relations. Data Min. Knowl. Discov. **26**(3), 574–619 (2013)
26. Cerf, L., Besson, J., Robardet, C., Boulicaut, J.-F.: Data peeler: constraint-based closed pattern mining in n-ary relations. In: Proceedings of the SIAM International Conference on Data Mining, SDM 2008, 24–26 April 2008, Atlanta, Georgia, USA, pp. 37–48. SIAM (2008)
27. Cerf, L., Besson, J., Robardet, C., Boulicaut, J.-F.: Closed patterns meet n-ary relations. ACM Trans. Knowl. Discov. Data **3**(1), 1–36 (2009)
28. Chataing, T., Perez, J., Plantevit, M., Robardet, C.: DiffVersify: a scalable approach to differentiable pattern mining with coverage regularization. In: Bifet, A., Davis, J., Krilavičius, T., Kull, M., Ntoutsi, E., Žliobaitė, I. (eds) ECML PKDD 2024. LNCS, vol 14946, pp. 407–422. Springer, Cham (2024). https://doi.org/10.1007/978-3-031-70365-2_24
29. Chen, Z., Chen, L., Villar, S., Bruna, J.: Can graph neural networks count substructures? In: Larochelle, H., Ranzato, M.A., Hadsell, R., Balcan, M.-F., Lin, H.-T. (eds.) Advances in Neural Information Processing Systems 33: Annual Conference on Neural Information Processing Systems 2020, NeurIPS 2020, 6–12 December 2020, Virtual (2020)
30. Dzeroski, S., Goethals, B., Panov, P. (eds.) Inductive Databases and Constraint-Based Data Mining. Springer (2010)
31. Fey, M., Lenssen, J.E., Morris, C., Masci, J., Kriege, N.M.: Deep graph matching consensus. In: 8th International Conference on Learning Representations, ICLR 2020, Addis Ababa, Ethiopia, 26–30 April 2020. OpenReview.net (2020)
32. Fischer, J., Vreeken, J.: Differentiable pattern set mining. In: Zhu, F., Ooi, B.C., Miao, C. (eds.) KDD 2021: The 27th ACM SIGKDD Conference on Knowledge Discovery and Data Mining, Virtual Event, Singapore, 14–18 August 2021, pp. 383–392. ACM (2021)
33. Fürnkranz, J., Knobbe, A.J.: Guest editorial: global modeling using local patterns. Data Min. Knowl. Discov. **21**(1), 1–8 (2010)
34. Galbrun, E., Cellier, P., Tatti, N., Termier, A., Crémilleux, B.: Mining periodic patterns with a MDL criterion. In: Berlingerio, M., Bonchi, F., Gärtner, T., Hurley, N., Ifrim, G. (eds.) ECML PKDD 2018, Part II. LNCS (LNAI), vol. 11052, pp. 535–551. Springer, Cham (2019). https://doi.org/10.1007/978-3-030-10928-8_32
35. Geerts, F., Goethals, B., Mielikäinen, T.: Tiling databases. In: Suzuki, E., Arikawa, S. (eds.) DS 2004. LNCS (LNAI), vol. 3245, pp. 278–289. Springer, Heidelberg (2004). https://doi.org/10.1007/978-3-540-30214-8_22

36. Georgii, E., Tsuda, K., Schölkopf, B.: Multi-way set enumeration in weight tensors. Mach. Learn. **82**(2), 123–155 (2011)
37. Guo, M., Chou, E., Huang, D.-A., Song, S., Yeung, S., Fei-Fei, L.: Neural graph matching networks for fewshot 3D action recognition. In: Ferrari, V., Hebert, M., Sminchisescu, C., Weiss, Y. (eds.) ECCV 2018, Part I. LNCS, vol. 11205, pp. 673–689. Springer, Cham (2018). https://doi.org/10.1007/978-3-030-01246-5_40
38. Imielinski, T., Mannila, H.: A database perspective on knowledge discovery. Commun. ACM **39**(11), 58–64 (1996)
39. Knobbe, A., Crémilleux, B., Fürnkranz, J., Scholz, M.: From local patterns to global models: the LeGo approach to data mining. LeGo **8**, 1–16 (2008)
40. Knobbe, A.J., Ho, E.K.Y.: Pattern teams. In: Fürnkranz, J., Scheffer, T., Spiliopoulou, M. (eds.) PKDD 2006. LNCS (LNAI), vol. 4213, pp. 577–584. Springer, Heidelberg (2006). https://doi.org/10.1007/11871637_58
41. Kontonasios, K.-N., Vreeken, J., De Bie, T.: Maximum entropy models for iteratively identifying subjectively interesting structure in real-valued data. In: Blockeel, H., Kersting, K., Nijssen, S., Železný, F. (eds.) ECML PKDD 2013. LNCS (LNAI), vol. 8189, pp. 256–271. Springer, Heidelberg (2013). https://doi.org/10.1007/978-3-642-40991-2_17
42. Lam, H.T., Moerchen, F., Fradkin, D., Calders, T.: Mining compressing sequential patterns. In: Proceedings of the Twelfth SIAM International Conference on Data Mining, Anaheim, California, USA, 26–28 April 2012, pp. 319–330. SIAM/Omnipress (2012)
43. Lam, H.T., Mörchen, F., Fradkin, D., Calders, T.: Mining compressing sequential patterns. Stat. Anal. Data Min. **7**(1), 34–52 (2014)
44. Li, Y., Gu, C., Dullien, T., Vinyals, O., Kohli, P.: Graph matching networks for learning the similarity of graph structured objects. In: Chaudhuri, K., Salakhutdinov, R. (eds.) Proceedings of the 36th International Conference on Machine Learning, ICML 2019, 9–15 June 2019, Long Beach, California, USA. Proceedings of Machine Learning Research, vol. 97, pp. 3835–3845. PMLR (2019)
45. Lijffijt, J., Kang, B., Duivesteijn, W., Puolamäki, K., Oikarinen, E., De Bie, T.: Subjectively interesting subgroup discovery on real-valued targets. In: 34th IEEE International Conference on Data Engineering, ICDE 2018, Paris, France, 16–19 April 2018, pp. 1352–1355. IEEE Computer Society (2018)
46. Lijffijt, J., Spyropoulou, E., Kang, B., De Bie, T.: P-N-RMiner: a generic framework for mining interesting structured relational patterns. Int. J. Data Sci. Anal. **1**(1), 61–76 (2016)
47. Liu, X., Pan, H., He, M., Song, Y., Jiang, X., Shang, L.: Neural subgraph isomorphism counting. In: Gupta, R., Liu, Y., Tang, J., Prakash, B.A. (eds.) KDD 2020: The 26th ACM SIGKDD Conference on Knowledge Discovery and Data Mining, Virtual Event, CA, USA, 23–27 August 2020, pp. 1959–1969. ACM (2020)
48. Livanos, M.J., Davidson, I.: Identification and uses of deep learning backbones via pattern mining. In: Shekhar, S., Papalexakis, V., Gao, J., Jiang, Z., Riondato, M. (eds.) Proceedings of the 2024 SIAM International Conference on Data Mining, SDM 2024, Houston, TX, USA, 18–20 April 2024, pp. 697–705. SIAM (2024)
49. Mannila, H., Toivonen, H.: Levelwise search and borders of theories in knowledgediscovery. Data Min. Knowl. Disc. **1**(3), 241–258 (1997)
50. Meo, R., Lanzi, P.L., Klemettinen, M. (eds.): Database Support for Data Mining Applications: Discovering Knowledge with Inductive Queries. Lecture Notes in Computer Science, vol. 2682. Springer (2004)

51. Morik, K., Boulicaut, J.-F., Siebes, A. (eds.) Local Pattern Detection, International Seminar, Dagstuhl Castle, Germany, April 12-16, 2004, Revised Selected Papers. Lecture Notes in Computer Science, vol. 3539. Springer (2005)

52. Morishita, S., Sese, J.: Traversing itemset lattice with statistical metric pruning. In: Vianu, V., Gottlob, G. (eds.) Proceedings of the Nineteenth ACM SIGMOD-SIGACT-SIGART Symposium on Principles of Database Systems, May 15-17, 2000, Dallas, Texas, USA, pp. 226–236. ACM (2000)

53. Nanni, M., et al.: Give more data, awareness and control to individual citizens, and they will help COVID-19 containment. Ethics Inf. Technol. **23**(S1), 1–6 (2021)

54. Ouali, A., et al.: Integer linear programming for pattern set mining; with an application to tiling. In: Kim, J., Shim, K., Cao, L., Lee, J.-G., Lin, X., Moon, Y.-S. (eds.) PAKDD 2017. LNCS (LNAI), vol. 10235, pp. 286–299. Springer, Cham (2017). https://doi.org/10.1007/978-3-319-57529-2_23

55. Pastor, E., Baralis, E.: Explaining black box models by means of local rules. In: Hung, C.-C., Papadopoulos, G. A. (eds.) Proceedings of the 34th ACM/SIGAPP Symposium on Applied Computing, SAC 2019, Limassol, Cyprus, 8–12 April 2019, pp. 510–517. ACM (2019)

56. Pastor, E., de Alfaro, L., Baralis, E.: Looking for trouble: analyzing classifier behavior via pattern divergence. In: Li, G., Li, Z., Idreos, S., Srivastava, D. (eds.) SIGMOD 2021: International Conference on Management of Data, Virtual Event, China, 20–25 June 2021, pp. 1400–1412. ACM (2021)

57. Pei, J., Han, J.: Can we push more constraints into frequent pattern mining? In: Ramakrishnan, R., Stolfo, S.J., Bayardo, R.J., Parsa, I. (eds.) Proceedings of the sixth ACM SIGKDD International Conference on Knowledge Discovery and Data Mining, Boston, MA, USA, 20–23 August 2000, pp. 350–354. ACM (2000)

58. Proença, H.M., van Leeuwen, M.: Interpretable multiclass classification by mdl-based rule lists. Inf. Sci. **512**, 1372–1393 (2020)

59. De Raedt, L., Zimmermann, A.: Constraint-based pattern set mining. In: Proceedings of the Seventh SIAM International Conference on Data Mining, 26–28 April 2007, Minneapolis, Minnesota, USA, pp. 237–248 (2007)

60. Shinji, T., Sugihara, R., Kitahara, R., Karasuyama, M.: Learning attributed graphlets: predictive graph mining by graphlets with trainable attribute. In: Baeza-Yates, R., Bonchi, F. (eds.) Proceedings of the 30th ACM SIGKDD Conference on Knowledge Discovery and Data Mining, KDD 2024, Barcelona, Spain, 25–29 August 2024, pp. 2830–2841. ACM (2024)

61. Siebes, A.: Data surveying: foundations of an inductive query language. In: Fayyad, U.M., Uthurusamy, R. (eds.) Proceedings of the First International Conference on Knowledge Discovery and Data Mining (KDD-95), Montreal, Canada, 20–21 August 1995, pp. 269–274. AAAI Press (1995)

62. Siebes, A., Vreeken, J., van Leeuwen, M.: Item sets that compress. In: Ghosh, J., Lambert, D., Skillicorn, D.B., Srivastava, J. (eds.) Proceedings of the Sixth SIAM International Conference on Data Mining, 20–22 April 2006, Bethesda, MD, USA, pp. 395–406. SIAM (2006)

63. Soulet, A., Crémilleux, B.: An efficient framework for mining flexible constraints. In: Ho, T.B., Cheung, D., Liu, H. (eds.) PAKDD 2005. LNCS (LNAI), vol. 3518, pp. 661–671. Springer, Heidelberg (2005). https://doi.org/10.1007/11430919_76

64. Soulet, A., Rioult, F., Crémilleux, B.: A condensed survey on condensed representations of patterns. In: Goethals, B., Robardet, C., A. (eds.) Proceedings of the 20th anniversary Workshop on Knowledge Discovery in Inductive Databases Co-located with the European Conference on Machine Learning and Principles and

Practice of Knowledge Discovery in Databases 2022 (ECMLPKDD 2022), Grenoble, France, 19–23 September 2022. CEUR Workshop Proceedings, vol. 3334, pp. 1–16. CEUR-WS.org (2022)

65. Spyropoulou, E., De Bie, T., Boley, M.: Interesting pattern mining in multirelational data. Data Min. Knowl. Discov. **28**(3), 808–849 (2014)

66. Uno, T.: An efficient algorithm for solving pseudo clique enumeration problem. Algorithmica **56**(1), 3–16 (2010)

67. Le Van, T., van Leeuwen, M., Nijssen, S., Fierro, A.C., Marchal, K., De Raedt, L.: Ranked tiling. In: Calders, T., Esposito, F., Hüllermeier, E., Meo, R. (eds.) ECML PKDD 2014. LNCS (LNAI), vol. 8725, pp. 98–113. Springer, Heidelberg (2014). https://doi.org/10.1007/978-3-662-44851-9_7

68. van Leeuwen, M., De Bie, T., Spyropoulou, E., Mesnage, C.: Subjective interestingness of subgraph patterns. Mach. Learn. **105**(1), 41–75 (2016). https://doi.org/10.1007/s10994-015-5539-3

69. van Leeuwen, M., Siebes, A.: STREAMKRIMP: detecting change in data streams. In: Daelemans, W., Goethals, B., Morik, K. (eds.) ECML PKDD 2008, Part I. LNCS (LNAI), vol. 5211, pp. 672–687. Springer, Heidelberg (2008). https://doi.org/10.1007/978-3-540-87479-9_62

70. van Leeuwen, M., Vreeken, J., Siebes, A.: Compression picks item sets that matter. In: Fürnkranz, J., Scheffer, T., Spiliopoulou, M. (eds.) PKDD 2006. LNCS (LNAI), vol. 4213, pp. 585–592. Springer, Heidelberg (2006). https://doi.org/10.1007/11871637_59

71. Vendrov, I., Kiros, R., Fidler, S., Urtasun, R.: Order-embeddings of images and language. In: Bengio, Y., LeCun, Y. (eds.) 4th International Conference on Learning Representations, ICLR 2016, San Juan, Puerto Rico, 2–4 May 2016. Conference Track Proceedings (2016)

72. Vespier, U., Knobbe, A., Nijssen, S., Vanschoren, J.: MDL-based analysis of time series at multiple time-scales. In: Flach, P.A., De Bie, T., Cristianini, N. (eds.) ECML PKDD 2012. LNCS (LNAI), vol. 7524, pp. 371–386. Springer, Heidelberg (2012). https://doi.org/10.1007/978-3-642-33486-3_24

73. Veyrin-Forrer, L., Kamal, A., Duffner, S., Plantevit, M., Robardet, C.: On GNN explainability with activation rules. Data Min. Knowl. Discov. **38**(5), 3227–3261 (2024)

74. Vreeken, J., Siebes, A.: Filling in the blanks - KRIMP minimisation for missing data. In: Proceedings of the 8th IEEE International Conference on Data Mining (ICDM 2008), 15–19 December 2008, Pisa, Italy, pp. 1067–1072. IEEE Computer Society (2008)

75. Vreeken, J., Mvan Leeuwen, A., Siebes, A.: Characterising the difference. In: Berkhin, P., Caruana, R., Wu, X. (eds.) Proceedings of the 13th ACM SIGKDD International Conference on Knowledge Discovery and Data Mining, San Jose, California, USA, 12–15 August 2007, pp. 765–774. ACM (2007)

76. Vreeken, J., van Leeuwen, M., Siebes, A.: KRIMP: mining itemsets that compress. Data Min. Knowl. Discov. **23**(1), 169–214 (2011)

77. Walter, N.P., Fischer, J., Vreeken, J.: Finding interpretable class-specific patterns through efficient neural search. In: Wooldridge, M.J., Dy, J.G., Natarajan, S. (eds.) Thirty-Eighth AAAI Conference on Artificial Intelligence, AAAI 2024, Thirty-Sixth Conference on Innovative Applications of Artificial Intelligence, IAAI 2024, Fourteenth Symposium on Educational Advances in Artificial Intelligence, EAAI 2014, 20–27 February 2024, Vancouver, Canada, pp. 9062–9070. AAAI Press (2024)

78. Ying, R., Fu, T., Wang, A., You, J., Wang, Y., Leskovec, J.: Representation learning for frequent subgraph mining. CoRR, abs/2402.14367 (2024)

The Minimum Description of Arno's Legacy in Rennes

Peggy Cellier[1,2,3] (iD), Sébastien Ferré[1,2] (iD), Elisa Fromont[1,2(✉)] (iD),
Luis Galárraga[1,2] (iD), and Alexandre Termier[1,2] (iD)

[1] Univ Rennes, IRISA - CNRS UMR 6074, Rennes, France
{peggy.cellier,sebastien.ferre,elisa.fromont,luis.galarraga,
alexandre.termier}@irisa.fr
[2] Inria, Rocquencourt, France
[3] INSA Rennes, Rennes, France

Abstract. We examine the influence of Arno Siebes's work on various data mining researchers in Rennes, France. A substantial aspect of this impact revolves around the Minimum Description Length Principle and its role in extracting a concise set of meaningful patterns from datasets. In Rennes, these patterns include (nested) itemsets, periodic itemsets, subsequences, and graphs. After briefly presenting the methods developed in our lab, we illustrate their application using datasets related to Arno's life and experiences.

1 Introduction

The IRISA/Inria lab in Rennes is one of the places where the seeds of MDL-based pattern mining planted by Arno Siebes [14] have flourished. However, these brave little seeds did not follow the simple Utrecht-Rennes route, and made quite a detour in time and space, getting nearly lost in the process.

The catalyst for Rennes' interest in the subject emerged in approximately 2010, when Alexandre Termier attended a talk by Arno Siebes at the LIG laboratory in Grenoble. The talk gave a very complete view on MDL-based pattern mining (for that time), by first presenting the Krimp algorithm [13], then demonstrating the relevance of the patterns found on classification tasks [11], and then showing how to use the approach for change detection in streams [10]. Alexandre was fascinated by how elegantly the MDL-based approach solved the long standing problem of getting **few** interesting patterns. So he immediately put it in the back of his head and forgot about it.

It may seem surprising in our enlightened era, but pre-2010, for many pattern miners, the real fun of pattern mining research was to make exhaustive algorithms faster and faster. Being "one order magnitude faster than the SOTA" was the way to get a good publication. Pattern interestingness was a nice idea when asking for project money... however in practice, when designing experiments for time and memory consumption, patterns were politely asked to go to /dev/null, so they did not get that much attention.

M. van Leeuwen and J. Vreeken (Eds.): Arno Siebes Festschrift, LNCS 16067, pp. 22–33, 2026.
https://doi.org/10.1007/978-3-032-03028-3_2

Better times arrived few years later. Alexandre had moved to Rennes, and from a scientific point of view, now most people (Alexandre included) had realized that making faster algorithms to send billions of patterns to `/dev/null` was not an extremely productive endeavour.

The tipping point to move to MDL approaches was a single remark made by Pr. Hiroki Arimura at the defense of Patricia Lopez Cuevas (a PhD student of Alexandre's) in 2013. She was working on mining periodic patterns in execution traces of embedded systems, and Hiroki said that "periodic patterns seem to be an excellent way to *compress the data*". Later in Rennes, when Alexandre was looking for a way to find "actually interesting" periodic patterns, he put together the talk of Arno and the remark of Hiroki, and knew he might be onto something. A good test for a new idea is to see if it survives the "whiteboard test" with other like-minded colleagues. A series of intense brainstorming sessions started with Peggy Cellier (also at IRISA) and Bruno Crémilleux (from Caen), soon joined by Esther Galbrun who had learned about the basics of MDL-based pattern mining through a collaboration with Matthijs van Leeuwen. Learning MDL took time, as well as help and patience from MDL evangelists. For example, Alexandre fondly remembers a face-to-face 2 h lesson on MDL that he received from Jilles Vreeken during KDD'17. Some words of this lesson still linger, such as *"Imagine the perfect encoding for your problem, designed by Grünwald himself"*...

This team, helped at the end by Nikolaj Tatti, proposed a first MDL-based periodic pattern mining approach [8], that will be presented in Sect. 2.1. From then on, this research topic branched out: Peggy started working with her colleague Sébastien Ferré on bringing MDL to the world of (knowledge) graphs (see Sect. 2.2). Meanwhile, Esther and Alexandre teamed up with Elisa Fromont and Luis Galarraga to study the discovery of MDL-based sequential rules, a work presented in Sect. 2.3.

And as an unexpected development of this story, Sébastien Ferré decided recently that MDL had not to be limited to pattern mining problems, and he used it to tackle the ARC-AGI intelligence challenge. We are pleased to report that in the 1st edition of the ARCathon competition, in 2022, his MDL-based approach got the 4th place, proposing a frugal and elegant alternative to many solutions built on brute-force search, and later on large language models. More details are provided in Sect. 2.4.

2 Methods

We now briefly summarize the different work done in our research institute (IRISA) in Rennes concerning pattern mining with the Minimum Description Length (MDL) principle.

The MDL principle [9] is a criterion rooted in information theory that stipulates that the best model for a dataset is the model that compresses it best. For a dataset D and a family of models $\mathcal{H}$, the best model for D, according to the (two-part) MDL, is the one that minimizes the description length

$$H^* = \arg \min_{H \in \mathcal{H}} \mathrm{L}(H) + \mathrm{L}(D|H), \qquad (1)$$

where L denotes code length in bits. Equation (1) strikes a balance between model complexity, as measured by L(H), and fitness to the data, as measured by L($D|H$). When applying MDL to pattern mining, a model is any collection of patterns. One of the main ingredients of any MDL-inspired pattern mining approach is the *encoding scheme*, namely, the protocol to encode the data with the patterns and to encode the patterns themselves. Once an encoding mechanism is in place, we can generate candidate sets of patterns, calculate the corresponding code lengths, and select the set resulting in the shortest code length.

In the following, we present the encoding schemes and sketch the algorithms developed in our lab for different types of patterns.

2.1   Periodic: Nested Patterns Sequences

Event logs are among the most ubiquitous types of data nowadays. They can be machine generated (server logs, database transactions, sensor data) or human generated (ranging from hospital records to life tracking, a.k.a. quantified self), and are bound to become ever more voluminous and diverse with the increasing digitisation of our lives and the advent of the Internet of Things (IoT). Such logs are often the most readily available sources of information on a system or process of interest. It is thus critical to have effective and efficient means to analyse them and extract the information they contain. Many such logs monitor repetitive processes, and some of this repetitiveness is recorded in the logs. A careful analysis of the logs can thus help understand the characteristics of the underlying recurrent phenomena. However, this is not an easy task: a log usually captures many different types of events. Events related to occurrences of different repetitive phenomena are often mixed together as well as with noise, and the different signals need to be disentangled to allow analysis. This can be done by a human expert having a good understanding of the domain and of the logging system, but is tedious and time consuming. In 2018, we presented a novel (at that time) approach for mining periodic patterns using a MDL criterion [8]. The main component of this approach–and our main contribution–was the definition of an expressive pattern language and the associated encoding scheme which allows to compute a MDL-based score for a given pattern collection and sequence. We designed an algorithm for putting this approach into practice and tested it on several event log datasets. We showed that we are able to extract sets of patterns that compress the input sequences and to identify meaningful patterns. However, it turned out that human activities and even machines are often not as perfectly regular as might have been anticipated, and that the ability of the approach to tolerate a deviations would need to be increased in order to accommodate real-world circumstances.

2.2   GraphMDL: Patterns in Graphs

Many fields have complex data that need labeled graphs, i.e., graphs in which both vertices and edges have labels. For example, in chemistry and biology, molecules are modeled as atoms connected by bonds; in linguistics, sentences are

P	G^P			c_P	v_π			c_π	
	Pattern structure	Pattern usage	Pattern code	Pattern code length (bits)	Port count	Port ID	Port usage	Port code	Port code length (bits)
P1	X —a— Y —b— Z ①②③	3	[P1]	1	2	v1 v2	1 3	[v1] [v2]	2 0.42
Pa	①—a—②	1	[Pa]	2.58	2	v1 v2	1 1	[v1] [v2]	1 1
Pw	①ʷ	1	[Pw]	2.58	1	v1	1		0
Px	①ˣ	1	[Px]	2.58	1	v1	1		0

Fig. 1. Example of a GraphMDL code table over the graph of Fig. 2. Pattern and port usages, and code lengths have been added as illustration and are not part of the table definition. Unused singleton patterns are omitted.

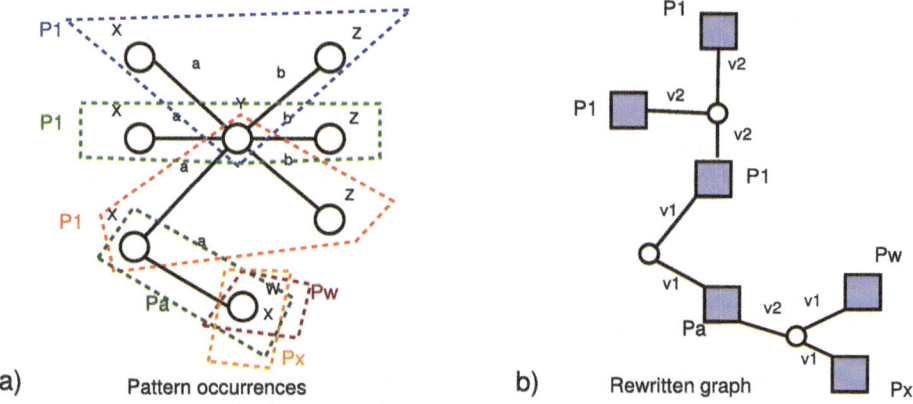

Fig. 2. *a)* Retained occurrences of patterns. *b)* The rewritten graph. Blue squares are pattern embeddings (their label indicates the pattern), white circles are port vertices. Edge labels represent which pattern port correspond to each port vertex.

represented by words linked through syntactic dependencies; and in the semantic web, knowledge graphs encode entities and their relationships. Numerous graph pattern mining algorithms, such as gSpan [15] and Gaston [12], have been developed to extract frequent substructures in such graphs. However, a major limitation of these approaches is that they often produce an overwhelming number of patterns, making human analysis difficult.

In [1–3], we proposed MDL-based approaches for selecting a characteristic subset of patterns on labeled graphs. The key idea of these approaches is the concept of ports, which capture how pattern occurrences connect to one another, without any loss of information. Ports are defined as a subset of a pattern's vertices: a vertex becomes a port if at least one embedding maps it to a vertex in the original graph that is also shared by another embedding–either of the same pattern or a different one. As a consequence we have to take into account of those ports in the encoding and thus in the code table. Consider the graph in Fig. 2a, where three instances of pattern $P1$ ($X-a-Y-b-Z$) overlap through their middle vertex labeled Y. This shared vertex is a port. The core intuition behind our approach is to represent the original graph as a set of interconnected pattern occurrences, linked via ports. Encoding the data with a code table (CT) involves building a structure, the rewritten graph, that explicitly shows which pattern embeddings are used and how they interconnect to reconstruct the original graph. Figure 2b illustrates the rewritten version of the graph from Fig. 2a encoded with the CT of Fig. 1. In the code table, we not only list the patterns and their associated code lengths, but also the ports of each pattern and the number of times each port is used. For instance, pattern $P1$ has two ports: v_1, used once, and v_2, used three times. In the rewritten graph, pattern embeddings are represented as blue square vertices, while shared vertices (ports) appear as white circles. When a pattern embedding involves a port, it is connected to the corresponding port vertex, with the edge label indicating the port's identity within the pattern. For instance, the three embeddings of pattern $P1$ all share the same Y-labeled vertex, leading to connections to a common port vertex via port v_2 in the rewritten graph. Notably, since port usage increases the overall description length, our method tends to favor patterns with fewer ports.

Experimental results show that our approach drastically reduces the number of patterns. Moreover, the selected patterns have complex shapes and are representative of the data.

2.3   Cossu: Patterns in Long Sequences

Next, we turned our interest to understanding the underlying generation process for long sequences of symbolic events [4]. Specifically, we proposed COSSU (COmpact Sets of Sequential rUles), an algorithm to mine small and meaningful sets of sequential rules. The rules are selected using an MDL-inspired criterion that favors compactness and relies on a novel rule-based encoding scheme for sequences.

The encoding scheme is defined as follows: Given a sequence S of m elements seen so far, and a set of weighted rules $(A_R \Rightarrow C_R, w_R)$, such that A_R, C_R, and w_R are respectively the antecedent, the consequent, and the weight[1] of rule R, our goal is to compute the probability of the next element given this set of rules – denoted by $\mathcal{R}$. This amounts to computing a probability distribution over Σ. For each possible next element $\sigma \in \Sigma$, we sum the weights of the active rules that predict σ and divide by the sum of weights of all active rules:

$$P_{\mathcal{R},S}(\sigma) = \frac{\sum_{(R,j)\in\mathcal{A}_{S,\sigma}} w_R}{\sum_{(R,j)\in\mathcal{A}_S} w_R}$$

where $\mathcal{A}_S = \{(R,j), (R, w_R) \in \mathcal{R} \text{ s.t. active}(R, S, j)\}$, and $\mathcal{A}_{S,\sigma} = \{(R,j) \in \mathcal{A}_S \text{ s.t. predict}(R, S, j) = \sigma\}$. A full sequence S of length n is transmitted element by element. At each step m, we encode the next symbol $S[m]$ with a code word chosen according to the probability assigned to it based on $\mathcal{R}$ and the

---

[1] This is a joint readjusted confidence score that, in addition, is also cheap to compress.

portion of the sequence seen so far, $S[1, m-1]$. Hence, the overall code length for the full sequence is

$$L(S|\mathcal{R}) = \sum_{m \in [1,n]} -\log_2 \left(P_{\mathcal{R}, S[1, m-1]}(S[m]) \right). \tag{2}$$

Overall, the code length for a rule R is the sum of the lengths of the code words for the antecedent, the consequent and the weight, respectively:

$$L(R) = \left[L_{\mathbb{N}}(|A_R| + 1) + \sum_{\sigma \in A_R} -\log_2(f_\sigma) \right]$$

$$+ \left[L_{\mathbb{N}}(|C_R|) + \sum_{\sigma \in C_R} -\log_2(f_\sigma) \right] + L_D(w_R).$$

We encode a rule table by stating the number of rules and then listing them

$$L(\mathcal{R}) = L_{\mathbb{N}}(|\mathcal{R}|) + \sum_{R \in \mathcal{R}} L(R). \tag{3}$$

Given a sequence S over alphabet Σ as input, COSSU returns a set of rules $\mathcal{R}$ in two phases. First, *rule construction* generates a collection of candidate rules $\mathcal{C}$, which are then evaluated during *rule selection*. The candidate rules can be mined with any standard SOTA rule miner on sequences, e.g., [6]. The selection phase starts with a set of $\mathcal{R}$ of background rules $\emptyset \Rightarrow \sigma$ (one for each $\sigma \in \Sigma$), for which we compute a baseline description length. At each step COSSU adds a candidate rule $R \in \mathcal{C}$ to $\mathcal{R}$, optimizes the weights of the rules, and keeps the rule if its addition reduces the compression length of S. An additional routine removes from $\mathcal{R}$ previously added rules that may have become superfluous due to the addition of R. After having processed all the candidates in $\mathcal{C}$, the algorithm reports the set of rules $\mathcal{R}$.

The rules extracted by COSSU constitute an interpretable model that exhibits competitive accuracy for the tasks of next-element prediction and classification.

2.4   MADIL-ARC: Patterns in Colored Grids

The Abstraction and Reasoning Corpus (ARC-AGI) is a benchmark that was introduced to foster AI research beyond narrow generalization [5]. It takes the form of a psychometric test where the agent, human or machine, has to learn a task from a few examples only. Each task is almost unique so that solving a task does not help much to solve another task. ARC-AGI tasks consist in transforming colored grids. Figure 3 shows an example task from the 800 public ARC-AGI tasks. Each demonstration example is made of an input grid (top) and an output grid (bottom). There are in general 2 or 3 examples (left of the red line). The agent has to predict the output grid of the test instance (right of

Fig. 3. Task 47c1f68c (inputs at the top, outputs at the bottom, test on the right)

the red line), given the input grid. The predicted grid must be identical to the expected grid to be successful. Most existing approaches rely either on brute-force search over a space of domain-specific programs, or on massively trained and fine-tuned LLMs based on various textual representations of colored grids.

We have proposed an MDL-based and pattern-based approach, called MADIL [7], that strives to compress the input grids on one hand, and the output grids on the other hand, as much as possible. This compression is conducted with nested patterns that decompose the grids into smaller and smaller parts. The hypothesis is that, if the correct decomposition is found, it becomes almost trivial to map the input parts to the output parts, and hence to enable the prediction of the output grid from the input grid. On the above example task, the decomposition works along the following model.

```
BgColor(bgcolor: black,
    Metagrid(sepcolor: ?,
        [ [ Monocolor(color: ?, mask: ?), Empty(size2: ?) ],
          [ Empty(size3: ?),                Empty(size4: ?) ] ]))
=>
BgColor(bgcolor = bgcolor,
    SymmetryHV(
        Monocolor(color = sepcolor, mask = mask)))
```

The input grid has a black background and its contents can be split into four subgrids and some separator color (e.g., red in first example). The top-left subgrid can be decomposed into a color (e.g., blue in first example) and a mask. The output grid has the same background color, and its contents can be folded two times by symmetry onto its top-left quadrant (SymmetryHV). This subgrid can in turn be decomposed into a color (the separator color from the input) and a mask (the same as in the input).

The encoding scheme follows two-part MDL with $L(M, D) = L(M) + L(D \mid M)$, where M is a task model and D is the set of training examples. Encoding a set of examples amounts to encoding each example. Encoding an example (v^i, v^o) (a pair of values, here grids) according to a model M amounts to encode a description $d = (d^i, d^o)$ of the example, where a description is a grounding of the model, replacing unknowns (?) and expressions by concrete values. A difficulty is that there may be several descriptions for an example. According to the MDL principle, we choose the most compressive description.

$$d_M^*(v^i, v^o) := arg\,min_{d \in M(v^i, v^o)} \operatorname{L}(d \mid M)$$

From there, the description length of data can be defined as follows.

$$\operatorname{L}(D \mid M) := \sum_{(v^i, v^o) \in D} \operatorname{L}(d_M^*(v^i, v^o) \mid M)$$

Encoding an example description can be decomposed into two parts: the input and the output, where the output model is parameterized by an *environment* σ mapping input variables to values.

$$\operatorname{L}(d \mid M) := \operatorname{L}(d^i \mid m^i) + \operatorname{L}(d^o \mid m^o[\sigma_{d^i}])$$

Therefore, it remains to define $\operatorname{L}(d \mid m[\sigma])$, the description length of a value when decomposed into a description d according to the model m applied to the given environment σ. This is done by induction on the tree structure common to the description and the model. Encoding an unknown value is simply encoding the value. Encoding an expression value is not necessary because the value can be computed from the expression and the provided environment. Encoding a composite value $d = P(d_1, \ldots, d_k)$ – where P is a pattern – can be reduced to encoding its parts because the whole value can be computed in a deterministic way from the values of the parts.

$$\operatorname{L}(d \mid m[\sigma]) := \sum_{i=1}^{k} \operatorname{L}(d_i \mid m_i[\sigma])$$

In summary, the encoding of an example amounts to the encoding of the atomic parts that cannot be computed from the input parts.

Task models are learned by conducting a Monte Carlo Tree Search over the model space. The root model is ? => ?, and the children models refine the current model by replacing an unknown by a pattern-based decomposition or a submodel of any kind by an expression. The refinements with higher compression gain have a higher prior. The MDL principle has proved very effective in finding the correct decomposition to the point that greedy search is sufficient in many cases to find a solution. Compared to brute-force search, it can therefore search deeper and reach more complex models. However, it is a challenging benchmark and MADIL only solves about 15% of the evaluation tasks, compared to 85% for a human. For comparison, a brute-force approach (Icecuber) achieves 31%, GPT-4o achieves 9%, and the recent reasoning model o3 – fine-tuned to ARC-AGI – achieves human level at 83% but at a very high inference cost.

3   Results

We now apply the methods outlined in Sects. 2.3 and 2.4 to novel datasets related to Arno Siebes' biography, and report the sometimes inconclusive findings that ensued.

3.1   ARC-AGI Tasks About Arno's Life Items

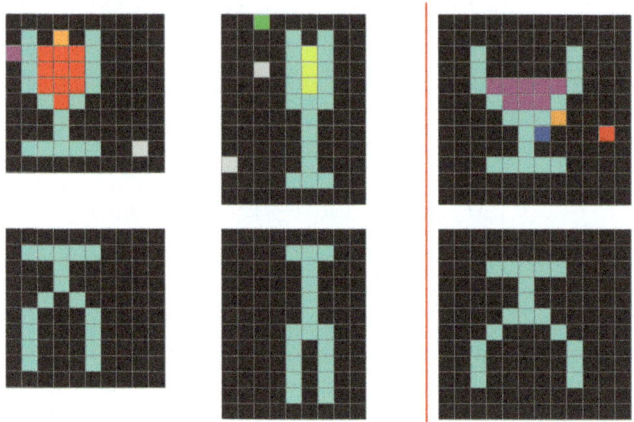

Fig. 4. ARC-Arno-Wine task about glasses of wine

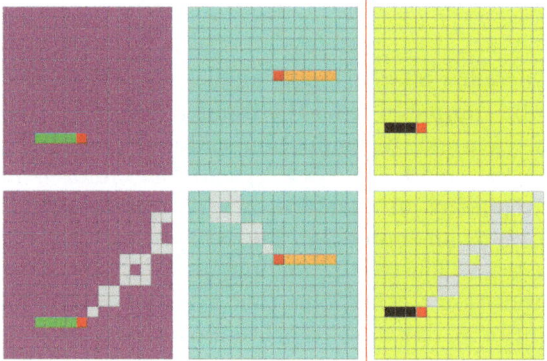

Fig. 5. ARC-Arno-Smoke task about smoke circles

We have designed three new ARC-AGI tasks reflecting Arno Siebes' life, shown in Figs. 4, 5, and 6. Let's see if the MADIL approach can solve them. For the first task, the following model is found.

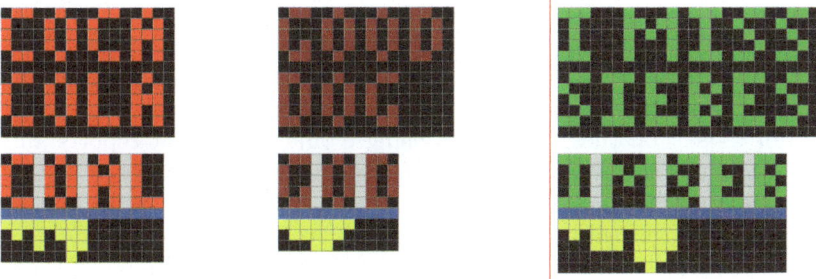

Fig. 6. ARC-Arno-MDL task about two-part MDL and play of words

```
BgColor(?bgcolor,
    Objects(?size, ?n, desc-area, Obj(?pos,
        Monocolor(?color, ?mask))))
=>
BgColor(black,
    Objects(!size, 1, ?, Obj(!pos[0],
        Monocolor(cyan, flipHeight(!mask)[0])))))
```

It reads: the input grid is made of a sequence of one-color objects over some background color, in descending area, and the output grid is made of a single cyan object at the same position as the first input object, and whose mask is the vertical flip of the mask of the first input object. Another explanation could be: "empty your glass, and turn it upside down!" The current state of MADIL fails to solve the two other tasks but we can see how it could in principle. In the task of Fig. 5, the grey object in the output could be decomposed into a sequence of squares with increasing sizes (smoke rings), chained diagonally upward, starting from the red point, and opposite to the colored segment next to the red point. The missing ingredients are relative object positions, and conditional submodels. The task in Fig. 6 is more complex as the output grid is a two-part MDL encoding of the input grid! The input is a sequence of objects (letters). The output can be decomposed into an upper part and a lower part. The upper part (code table) can be further decomposed into the sequence of unique letters from the input, and the lower part can be decomposed into a sequence of integers that encode the index of the input letters in the upper part. The last example reveals the author of the ARC-Arno tasks!

3.2   COSSU on Arno's Publication History

Publication Chronology. We downloaded Arno's publication chronology from DBLP and constructed three event sequences of length 74 from this information. The *venue sequence* encoded the names of the conferences of Arno's published papers as events. Given the diversity of symbols in the alphabet of this sequence—$|\Sigma| = 20$, which makes recurrent patterns unlikely to emerge—, we reduced the number of symbols by mapping the venues to topics, e.g., databases, data mining. This resulted in a second sequence, the *topic sequence*, that contained $|\Sigma| = 8$ different symbols. A third sequence used the last names of the first authors of Arno's papers, which resulted in a sequence with $|\Sigma| = 26$ different symbols that we call the *first-author sequence*.

We ran COSSU on these sequences using a rule miner that reports all rules $A \Rightarrow C$ with at least 2 occurrences in the sequence. In all cases COSSU found rule sets $\mathcal{R}$ containing only the background rules, i.e., of the form $\emptyset \Rightarrow \sigma$. An inspection at the algorithm's execution trace suggested that COSSU did find rules, e.g., *Siebes* $\Rightarrow$ *Siebes, Siebes* for the first-author sequence, but none of them was frequent and confident enough to reduce the baseline description length. This is presumably due to the significant variety of topics, venues, and collaborations of Arno's throughout his career.

Analysis of the Krimp Paper. We ran COSSU on the sequence of words of [14], one of the most representative works of Arno. After removing all stopwords, we obtained a sequence of length 10232 and an alphabet Σ of 2025 symbols. COSSU found two non-trivial rules that compressed the sequence: $et \Rightarrow al$ (64 occurrences, 100% confidence) and $code \Rightarrow table$ (148 occurrences, 59% confidence). An interesting question would be to check whether these rules (and some others) are characteristic of MDL papers.

4 Conclusion

This paper aims to pay tribute to the significant impact of Arno Siebes' work on the worldwide data mining community, with a special emphasis on its influence within our French cluster. Our work showcases various applications of the MDL principle in pattern mining, an area heavily inspired by Arno's contributions, including nested itemsets, periodic itemsets, subsequences, and graph-based approaches. Notably, the current progress on the ARC-AGI problem holds significant potential for developing more human-like artificial intelligence and may continue Arno's legacy for years to come.

References

1. Bariatti, F., Cellier, P., Ferré, S.: GraphMDL: graph pattern selection based on minimum description length. In: Berthold, M.R., Feelders, A., Krempl, G. (eds.) IDA 2020. LNCS, vol. 12080, pp. 54–66. Springer, Cham (2020). https://doi.org/10.1007/978-3-030-44584-3_5
2. Bariatti, F., Cellier, P., Ferré, S.: GraphMDL+: interleaving the generation and MDL-based selection of graph patterns. In: ACM/SIGAPP Symposium on Applied Computing (SAC), pp. 355–363. ACM (2021)
3. Bariatti, F., Cellier, P., Ferré, S.: KG-MDL: mining graph patterns in knowledge graphs with the MDL principle. CoRR **abs/2309.12908** (2023)
4. Bourrand, E., Galárraga, L., Galbrun, E., Fromont, É., Termier, A.: Discovering useful compact sets of sequential rules in a long sequence. In: 33rd IEEE International Conference on Tools with Artificial Intelligence, ICTAI, pp. 1295–1299. IEEE (2021)

5. Chollet, F.: On the measure of intelligence. arXiv:1911.01547 (2019)
6. Cule, B., Goethals, B.: Mining association rules in long sequences. In: Advances in Knowledge Discovery and Data Mining, pp. 300–309 (2010)
7. Ferré, S.: Tackling the abstraction and reasoning corpus (ARC) with object-centric models and the MDL principle. In: Miliou, I., Piatkowski, N., Papapetrou, P. (eds.) IDA 2024. LNCS, vol. 14641, pp. 3–15. Springer, Cham (2024). https://doi.org/10.1007/978-3-031-58547-0_1
8. Galbrun, E., Cellier, P., Tatti, N., Termier, A., Crémilleux, B.: Mining periodic patterns with a MDL criterion. In: Berlingerio, M., Bonchi, F., Gärtner, T., Hurley, N., Ifrim, G. (eds.) ECML PKDD 2018. LNCS (LNAI), vol. 11052, pp. 535–551. Springer, Cham (2019). https://doi.org/10.1007/978-3-030-10928-8_32
9. Grünwald, P.: The Minimum Description Length Principle. The MIT Press, Cambridge (2007)
10. van Leeuwen, M., Siebes, A.: STREAMKRIMP: detecting change in data streams. In: Daelemans, W., Goethals, B., Morik, K. (eds.) ECML PKDD 2008. LNCS (LNAI), vol. 5211, pp. 672–687. Springer, Heidelberg (2008). https://doi.org/10.1007/978-3-540-87479-9_62
11. van Leeuwen, M., Vreeken, J., Siebes, A.: Compression picks item sets that matter. In: Fürnkranz, J., Scheffer, T., Spiliopoulou, M. (eds.) PKDD 2006. LNCS (LNAI), vol. 4213, pp. 585–592. Springer, Heidelberg (2006). https://doi.org/10.1007/11871637_59
12. Nijssen, S., Kok, J.N.: The Gaston tool for frequent subgraph mining. Electron. Notes Theor. Comput. Sci. **127**(1), 77–87 (2005)
13. Siebes, A., Vreeken, J., van Leeuwen, M.: Item sets that compress. In: Proceedings of the Sixth SIAM International Conference on Data Mining, pp. 395–406. SIAM (2006)
14. Vreeken, J., van Leeuwen, M., Siebes, A.: Krimp: mining itemsets that compress. Data Min. Knowl. Discov. **23**(1), 169–214 (2011)
15. Yan, X., Han, J.: gSpan: graph-based substructure pattern mining. In: Proceedings of the 2002 IEEE International Conference on Data Mining (ICDM 2002), pp. 721–724. IEEE Computer Society (2002)

Did this Leopard Change his Spots?
Arno Siebes' Publication Trends Found Using Pattern Mining and Itemsets that (Don't) Compress (Very Well)

Bruno Crémilleux[1]([✉])(iD), Justine Reynaud[1](iD), Arnaud Soulet[2](iD),
Athénaïs Vaginay[1](iD), and Albrecht Zimmermann[1](iD)

[1] Université Caen Normandie, ENSICAEN, CNRS, Normandie Univ,
GREYC UMR 6072, 14000 Caen, France
`bruno.cremilleux@unicaen.fr`
[2] Université de Tours, LIFAT, Blois, France

Abstract. On the occasion of Arno Siebes' retirement, we revisit pattern mining to explore Arno-centric bibliographic data and highlight some trends in the output of his scientific career. By using pattern condensed representations and redescription mining, we shed new light on Arno's work and retrace his career. Interestingly, we show that Arno's data do not compress well, reflecting a particularly varied publication activity where redundancy is limited. Finally, we explore bibliographic data of his most frequent co-authors, Matthijs van Leeuwen and Jilles Vreeken, former PhD students of his, in order to determine if they stay in the lane that they explored under their supervisor, or branched out into new areas.

Keywords: Arno Siebes co-authorship · condensed representations of patterns · redescription mining · concept lattice · itemset compression · Krimp · MDL

1 Introduction

Analysing bibliographic data of researchers is a popular topic [9,16]—particularly once near the end of their career—to understand and highlight their collaborations, preferred venues, and research topics, and changes thereof. In addition, we argue that something even more interesting might be how certain authors' work differs (or not) from the output of the people they worked with or mentored.

In US sports, there is the concept of a "coaching tree": assistant coaches learning under a head coach then often use similar concepts when they become head coaches themselves [28]. In academics, PhD students and post-docs can be thought of as assistant coaches, and those that survive the academic gauntlet eventually become primary investigators themselves. At which point the question

---

The authors are listed in alphabetical order and contributed equally.

M. van Leeuwen and J. Vreeken (Eds.): Arno Siebes Festschrift, LNCS 16067, pp. 34–55, 2026.
https://doi.org/10.1007/978-3-032-03028-3_3

becomes: *Do they stay in the lane that they explored under their supervisor, or do they branch out into new areas?*

Some differences can be spotted quickly: Jilles Vreeken, to pick a completely random example, has publications in NeurIPS (formerly NIPS), which is not the case for Arno. Matthijs van Leeuwen[1] has clearly health-targeted publications, which cannot be found on his mentor's publication list. To go deeper, however, and understand whether concepts changed, we will need to look at publication content to a certain degree, which is where pattern mining comes in.

Proponents of pattern mining[2] claim that methods developed in this field, e.g. frequent (closed) itemset, pattern set, and redescription mining, can be used to derive interesting, interpretable insights into data. Especially in the context of this liber amicorum, we aim for descriptions of the person being lauded, instead of the predictive power of more complicated black box models. We will therefore use this chapter to put our claim to the test.

Hence, we investigate the following research questions (RQ): (RQ1) What characterizes Arno's publications policy according to the co-authorship, the topics, the periods? Can we highlight shifts and persistent themes in Arno's scientific career? (RQ2) Can Arno's bibliographic data be compressed? Data compression is an indicator of the variety of the data (the better bibliographic data can be compressed, the more uniform the publication policy with little variety of themes and/or co-authors) (RQ3) Mentor vs mentees dynamics: do Jilles and/or Matthijs publish like Arno? (or vice versa).

To answer RQ1, we start by studying the publication policy of Arno according to data available in the DBLP database. We do not restrict ourselves to Arno's co-authors, but also look at the publication dates and venues (conferences, journals), as well as the topics of the articles via their keywords. Our analysis is based on pattern summarizations in the form of Galois lattices and redescriptions. Whereas local patterns highlight associations or correlations in data, redescriptions identify sets of objects that admit multiple shared descriptions called views, each corresponding to different representations of the data [11]. We address RQ2 by using the Minimal Description Length principle (MDL) [21]. As first proposed by Arno, Jilles, and Matthijs [27], the MDL principle can be employed in pattern mining by looking for the set of frequent patterns that yields the best lossless compression of the data. Regarding RQ3, we compare Jilles' and Matthijs' publication data with those of Arno, in terms of closed patterns composed of words extracted from abstracts, as well as a subset selected using the MDL principle.

Taken as a whole, we hope that this work shows the potential of pattern mining for providing interpretable information in data analysis. Writing this chapter on the occasion of Arno's retirement in addition gives us an opportunity to shine a spotlight on our colleague's academic work.

This paper is structured as follows. Section 2, quickly introduces the notions and methods employed in this chapter. Section 3 details how we prepared the

---

[1] As another random example.

[2] Such as the authors of this chapter, even in the age of deep learning.

data we use. For each research question, Sect. 4 presents the experimental setup along with the conclusion we draw from the experiments. Finally, we discuss what we learnt and whether pattern mining proved to be as useful as we expected (Sect. 5).

2   A Crash Course in Pattern Mining

Pattern Mining and Redescription. Pattern mining is a central concept in knowledge discovery from databases, resulting in interpretable insights. The field is arguably firmly established with the publication of the paper introducting the APRIORI algorithm for frequent pattern mining [1]:

Definition 1. *Consider a set $\mathcal{I}$ of symbols called* items. *The multiset $\mathcal{T} = \{t_i \mid t_i \subseteq \mathcal{I}\}$ is called* transaction set. *Given an itemset $p \subseteq \mathcal{I}$, its* coverage *in $\mathcal{T}$ is $cov(p, \mathcal{T}) = \{t \in \mathcal{T} \mid p \subseteq t\}$, and its* support *$sup(p, \mathcal{T}) = |cov(p, \mathcal{T})|$ (relative support $sup(p, \mathcal{T}) = \frac{|cov(p, \mathcal{T})|}{|\mathcal{T}|}$). Given a threshold $\sigma \in \mathbb{N}$, p with $sup(p, \mathcal{T}) \geq \sigma$ is a* frequent itemset.

As we chose to describe the data used throughout this paper by items, we use the pattern language of itemsets in the definitions of this section. They can be generalized, however, to more complex pattern languages such as sequences [30], tree [31], or graphs [29].

While minimum frequency is the most powerful constraint when the goal is mining patterns *efficiently*, one can add other constraints [13,20] to fine-tune the information one wants to recover. Such patterns, evaluated individually and representing local phenomena, are typically referred to as *local patterns*, an area of research on which Arno co-organized influential Dagstuhl seminars [2,4].

The community quickly figured out that mining in such a way leads to too many results and proposed condensed representations. The key idea of pattern condensed representations is to exploit redundancies of a collection of patterns to construct a concise representation of the whole set instead of showing it in its entirety. Research on pattern condensed representations begins early on in pattern mining [3,6,19,25,32]. Here, we will only define closed and free itemsets, which are the representative patterns of condensed representations.

Definition 2. *An itemset p is called a* closed itemset *(resp. a* free itemset*) iff $\forall i \in \mathcal{I} \backslash p : sup(p \cup \{i\}, \mathcal{T}) < sup(p, \mathcal{T})$ (resp. $\forall i \in p : sup(p \backslash \{i\}, \mathcal{T}) > sup(p, \mathcal{T})$).*

Given the set of frequent closed/free itemsets, one can recover the support information of all frequent itemsets, hence the term *condensed*.

Closed and free patterns are arguably simplifications drawn from the field of *Formal Concept Analysis* [17], a field with which Arno is also well-acquainted [5]. Formal concepts can be grouped into *lattices* for visualization, something that we will employ later on.

Redescription mining [11], which will also use later, can also exploit these notions.

Definition 3. *Given two logical formulae X and Y such that $X \cap Y = \emptyset$ over items in a data set $\mathcal{T}$, we say that X and Y are redescriptions of the same subset $\mathcal{D} \subseteq \mathcal{T}$ if the Jaccard similarity of their respective coverages is greater-than-or-equal to a threshold σ_{rm}:*

$$J(X,Y) = \frac{|cov(X,\mathcal{T}) \cap cov(Y,\mathcal{T})|}{|cov(X,\mathcal{T}) \cup cov(Y,\mathcal{T})|} \geqslant \sigma_{rm}.$$

Interestingly, it is possible to show that two itemsets X and Y (such that $X \cap Y = \emptyset$) belonging to the same equivalence class constitute a redescription [33].

Even condensed representations are often redundant however, requiring a *global* view on pattern mining [14], which eventually leads to *pattern set mining* in a variety of ways, e.g. by using redundancy measures [15], encoding $\mathcal{T}$ using a subset of closed frequent itemsets [26] or explicitly enforcing other constraints (e.g. on the overlap in $\mathcal{T}$) on the set of patterns [8]. This view, which considers *subsets* of pattern condensed representations as *global models* rather than just a way of speeding up pattern mining, can be very helpful to tackle a number of tasks such as conceptual clustering [7,10], for instance.

Minimum Description Length. A particularly relevant pattern set mining example for this chapter is the seminal paper [26] that proposed using MDL [12] to choose the patterns that best compress the dataset. It formed the starting point for an important part of Arno's work that deals with dataset compression for selecting small but high-quality collections of patterns.

The size (i.e. number of patterns) of a pattern condensed representation gives an indication of the compression of the *output* of a mining process, but not of the compression of the *dataset*. Whereas the MDL-approach to pattern set mining [26] chooses the patterns that best compress the dataset and can be summarized as:

Definition 4. *Given a set of patterns $\mathcal{P}$, the best subset $\mathcal{P}_{opt} \subseteq \mathcal{L}$ is the one for which*

$$L(\mathcal{T}) = \min_{\mathcal{P}} L(\mathcal{P}) + L(\mathcal{T}|\mathcal{P}).$$

$L(\mathcal{P})$, and $L(\mathcal{T}|\mathcal{P})$ are adaptations of the MDL principle to local patterns. The former term represents the encoded length of the selected pattern set, called a *code table*, with individual patterns being assigned a code length in inverse proportion to their usage frequency in encoding $\mathcal{T}$. The latter term represents the encoded length of the transaction set, i.e. transactions expressed by a combination of patterns $p \in \mathcal{P}$, which in turn are translated into their respective codes.

The best code table thus represents a trade-off between representing the complexity of the underlying data and reduced encoding cost. This line of research is a corner stone of the doctoral dissertations of Jilles and Matthijs and also sparked more conceptual discussions [22].

Table 1. Number of publications by the person considered and the type of publication. Note that the datasets used in our analysis are derived only from *authored* publications.

Person	Authored			Edited
	Total	Published	Preprint	
Arno (1993–2024)	84	77	7	12
Jilles (2006–2024)	192	143	49	4
Matthijs (2006–2024)	95	72	23	3
At total	347			17

Table 2. Number of distinct attributes for each attribute type in each dataset. The year period corresponds to the temporal periods identified in Fig. 1.

Attribute type	# possible values for this attribute					
	Data-X-dblpK			Data-X-abstract		
	AS	JV	MvL	AS	JV	MvL
Author	100	102	73			
Venue	44	33	39			
Year period	4	2	3			
Keywords	317	494	342			
Terms				1407	2134	1578
Total (# of items)	465	631	457	1407	2134	1578

3   The Data and How We Prepared it

We consider any publication from the DBLP (Digital Bibliography & Library Project) database [18] where Arno, Matthijs, and/or Jilles are listed as either authors or editors. DBLP is a comprehensive bibliographic database of computer science publications. For each publication, we query DBLP using its SPARQL endpoint. The full SPARQL query can be found in Appendix A. It was run January 10, 2025 and returns the list of *authors*, *title*, *venue*, and *year* of publication for the 347 authored and 17 edited publications considered (cf. Table 1).

Table 1 summarises general information about the collected data. An interesting first observation is how publication volume/speed has changed over the years: over a 31-year period, Arno has published (and has had to publish) fewer papers than Matthijs over 18 years, and *far* fewer than Jilles.

Abstract and paper keywords are obtained either automatically, via web scraping, or manually from Semantic Scholar when scraping is not feasible. In the latter case, the keywords may not correspond to those explicitly stated in the original publication as they are auto-generated by Semantic Scholar's indexing system.

Abstracts and keywords then undergo a preprocessing step. We lemmatize the text of the abstracts and compute Term Frequency-Inverse Document Frequency

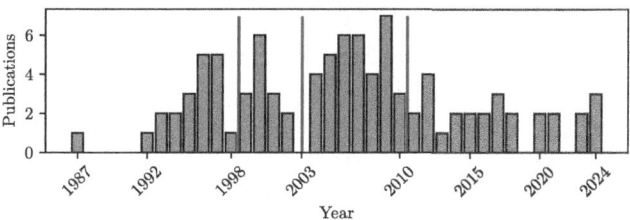

Fig. 1. Distribution of Arno's publications over time, which guides the definition of the four temporal periods used in subsequent analyses: 1987 1998, 1999 2003, 2004–2010, and 2011–2025.

(TF-IDF) scores. Stop words such as "be", "can", "do", "have", are used as a baseline to identify the *most relevant terms in the abstracts*: any term with a lower TF-IDF value than any of these stop words is discarded. Translating distinct terms into items to derive transactional data sets then leaves us with the number of items shown at the bottom of Table 2. As for the *publication keywords*, we apply a normalization step, by trimming trailing 's' from plural forms, improving consistency of the keywords across publications.

To support temporal analyses, we added an attribute to each publication indicating the period in which it was published. These periods were defined based on the distribution of Arno's publications (Fig. 1), resulting in the following four *periods*: 1987–1998, 1999–2003, 2004–2010, and 2011–2025.

For downstream analyses, we structure the authored publications into six datasets, grouped by author and by data type:

- DATA-AS-DBLPK, DATA-JV-DBLPK, DATA-MVL-DBLPK contain the metadata from DBLP (authors, title, venue, the date period—among the four we defined above), and the paper keywords, for Arno, Jilles and Matthijs respectively. Details about the attributes are given in the top part of Table 2.
- DATA-AS-ABSTRACT, DATA-JV-ABSTRACT, DATA-MVL-ABSTRACT contain the terms extracted automatically from the abstracts as depicted above. Numbers of terms and therefore number of items are given in the bottom part of Table 2.

The code for the data collection and preparation is available at https://git.unicaen.fr/justine.reynaud/liberamicorumas.

4 Experiments and Results

This section presents a series of exploratory analyses that aim at uncovering meaningful patterns within the publication data of Arno in order to highlight some trends of his scientific career. By using redescriptions and pattern summarizations, we adress RQ1 in Sects. 4.1 and 4.2. Applying the MDL principle, we answer RQ2 in Sect. 4.2. Finally, we consider RQ3 by studying the mentor vs mentees dynamics in Sect. 4.3.

4.1 Describing and Understanding Arno's Publications: Co-authorship Relations, Periods, Venues, and Topics

As a first exploration, we used redescription mining to understand Arno's publication history a bit better. The underlying datasets are DATA-AS-DBLPK, DATA-JV-DBLPK, DATA-MvL-DBLPK, we have placed the tables listing the resulting redescriptions in Appendix C to optimize the flow of the paper (Tables 6 to 7). For easier reading, we have grouped the redescriptions "thematically".

Usually, one would expect Jaccard values that are relatively high (≥ 0.8), which is not the case here. However, in this context, a Jaccard value of 0.4 for two authors indicates that 40% of their total publications are co-authored.

- Conditioned on Arno as a co-author (Table 6), for instance, Arno Knobbe[3] is virtually synonymous with ILP.
- Links for Arno with Jilles and Matthijs, as well as the ICDM conference, show up for the 2004–2010 period. A similar block is more strongly expressed in Jilles' data (Table 8), and links the three authors explicitly (we also find such a link as R8 in Table 7).
- Jilles has *a lot* more frequent co-author pairings than Arno and Matthijs, is the only one who has a clear "MDL" block, and has published on a number of topics at ICDM. That may be due to the fact that Jilles is more consistent in his choice of keywords.
- In Matthijs' data, we interestingly find a block describing his post-doctoral work with Luc De Raedt and Siegfried Nijssen that seems more strongly expressed, in terms of number of rules, than the one with Arno and Jilles. We think this is due to the number of items that co-occur: the more items there are, the more redescription we can find.

A Galois Lattice for Describing Arno's Data. A different way of looking at Arno's publication history consists of exploring the lattice of formal concepts shown in Fig. 2. The root of the lattice consists of the author item "Arno Siebes" given that it is present in all publications. As usual in such lattices, connecting lines denote superset/subset relations, e.g. the root item is included in all itemsets at the next level. The lattice can be decomposed into three broad regions:

- One involving the author's "early" period, characterized by the keywords association rule and data mining and publications at PKDD.
- One from 2004 to 2010, which groups publications with Jilles and Matthijs and makes the keyword MDL appear.
- And a third one ranging from 2011 to 2025. This period includes the first papers of Arno's uploaded to CORR in 2015[4], and publications with Roel Bertens.

Somewhat surprisingly, there is no concept grouping Arno, Jilles, Matthijs and MDL, which we suspect is due to the fact that the latter appears in several variants (followed or not by "principle", as acronym, or spelled out).

---

[3] Another of Arno Siebes' PhD students.
[4] Notably both written with Jilles.

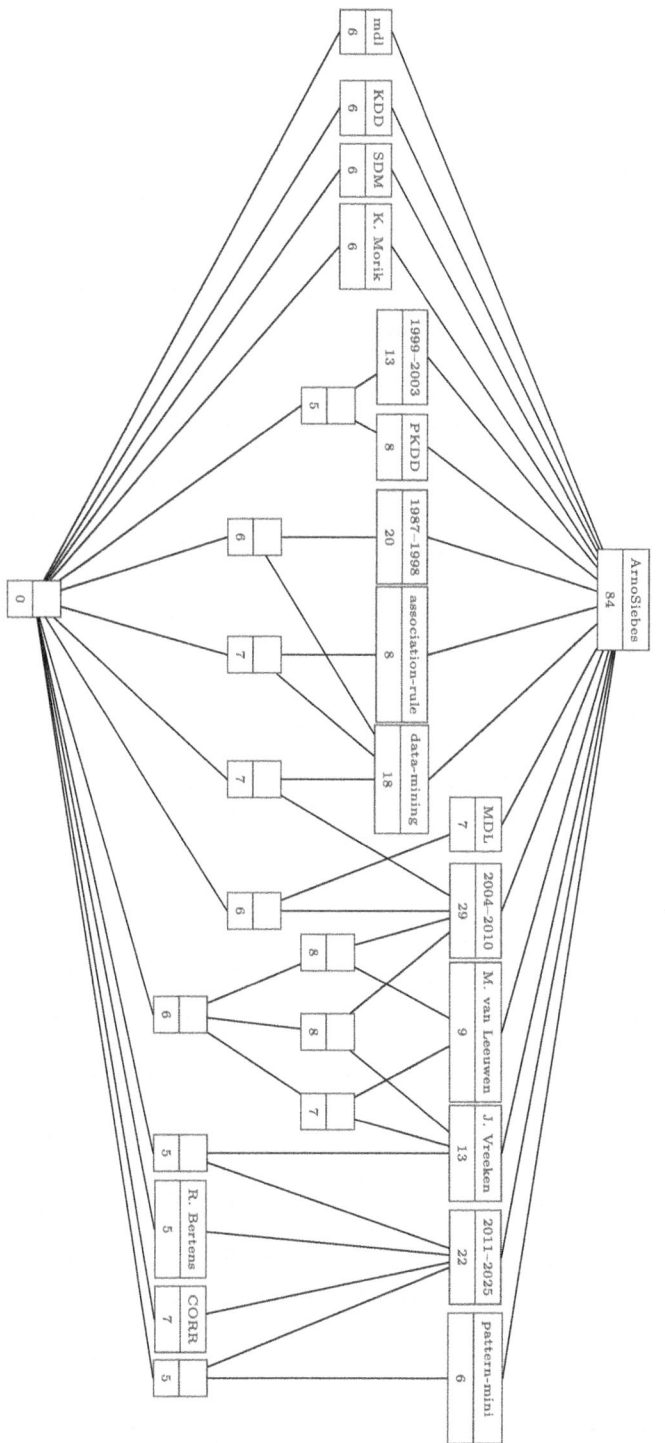

Fig. 2. Formal concept lattice derived from Arno's publication data ($\sigma = 5\%$'). In the lattice, "mdl" corresponds to the keyword "minimum-description-length".

4.2   Highlighting Trends in Arno's Scientific Career

Experimental Protocol and Presentation of Results. Using the SLIM algorithm [24], we applied the MDL principle to the set of itemsets having support greater than or equal to 2 in DATA-AS-DBLPK. Figure 3 shows the 39 patterns that make up the code table (excluding singletons)[5]. The upper part of the figure contains patterns including a year interval in their description. In the lower part, the patterns did not contain a year interval, but as a year is associated to each publication, by referring to the transactions encoded by a pattern, it is possible to approximately distribute each pattern over time.

Since patterns can contain items from 3 categories (authors, keywords and venues), they are broken down into these three distinct layers. A vertical line between several itemsets means that these itemsets form a single pattern in the code table. For example, pattern 33 is made up of author Thieme and keyword schema integration. Each pattern is also described by a pair of values indicating the pattern's coverage and frequency respectively. For example, pattern 33 is used only once to encode a transaction, whereas it is present 3 times in the dataset (in this case, pattern 5, which compresses better, is used the other 2 times). Finally, all the patterns in the upper part are simplified because they originally contain the author Siebes and an interval pattern according to the slice. For example, pattern 21 corresponds to Siebes, 1987–1998 and van der Voort.

Analysis of the Results. Figure 3 provides a few insights. Temporal dynamics are crucial, as 24 of the 39 patterns contain a time interval. Generally speaking, between 1987 and 1998, papers were mainly concerned with database issues at the conceptual level, as indicated by the keywords (schema integration, class hierarchy, query processing) or the main source of publications (CAiSE). Around 1999, we see a turning point with the keywords KDD and knowledge discovery (even though Arno was a pioneer, presenting himself as a data mining researcher as early as 1996 [23]), followed by the data mining wave of patterns 36, 37 and 10, which includes the ICDM conference. On the one hand, there is the work on classification trees (classification tree, classification tree analysis) and the work on propositionalization (with the ILP work of Arno Knobbe published in the PKDD conference). On the other hand, there is work on local pattern discovery mentioned by association rules over the period 1999–2010 and pattern mining over the period 2011–2025. Of course, the 2004–2010 interval (and even beyond) is strongly marked by work based on the MDL principle (cf. patterns 11, 16, 24) and work with Jilles and Matthijs, who appear in no less than 6 patterns of the code table (cf. patterns 4, 13, 18, 19, 34, 38) and 6 redescriptions (cf. R12, R18, R22, R21, R25, R6 in Table 6). Particular attention is also paid to sequential data using the MDL principle (pattern 3) and Markov chains (pattern 6). Figure 3 highlights ICDM and PKDD as venues (12 papers at ECML and PKDD). Arno is also a great contributor to SDM and KDD but these venues do

---

[5] Acronyms denoted by * are described in full in DATA-AS-DBLPK, but shortened for clarity.

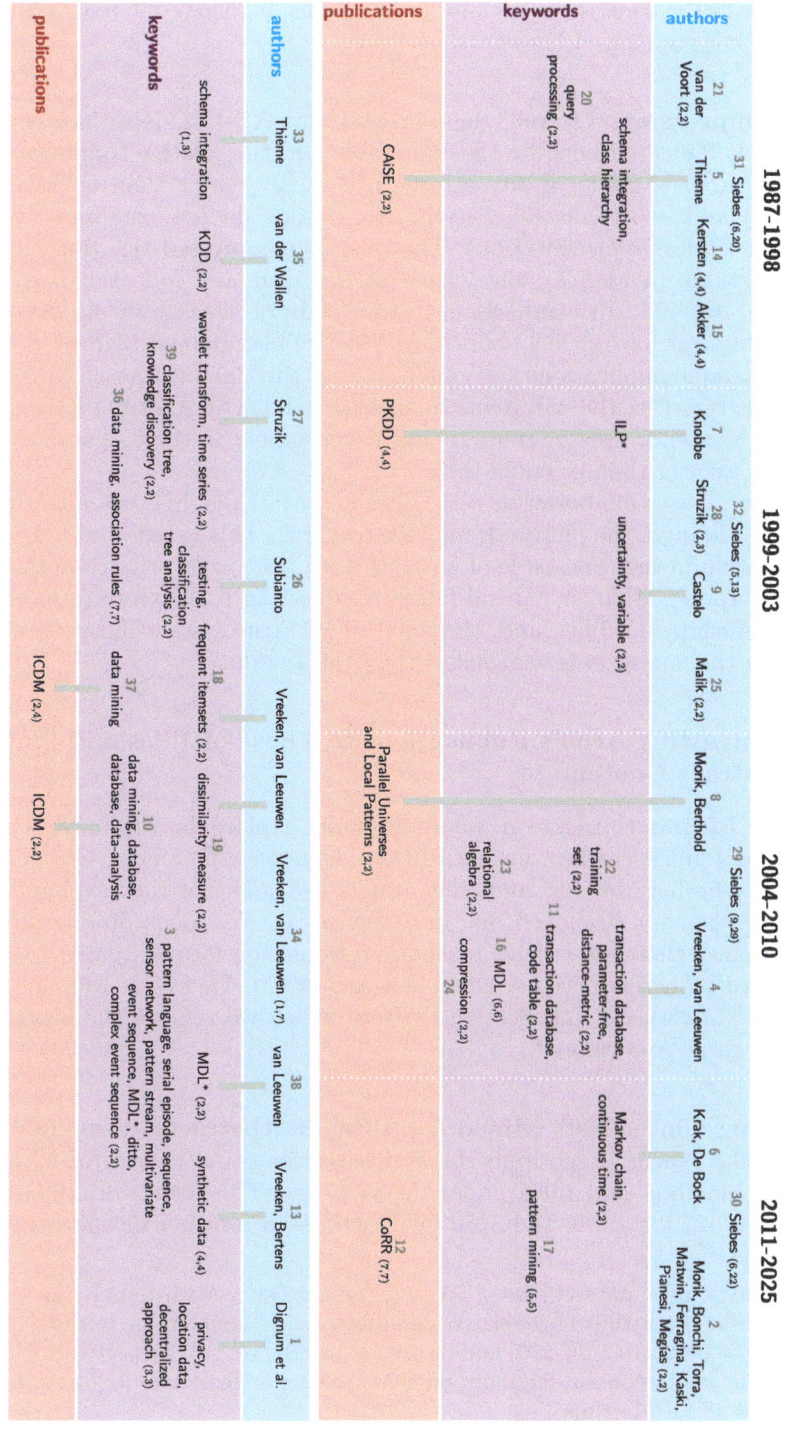

Fig. 3. Timeline of the code table for DATA-AS-DBLPK

not appear in the code table because publications at these conferences involve a variety of co-authors and keywords.

Data Compression. Overall, the dataset DATA-AS-DBLPK do not compress well. This is indicated both by the code table and the relative compressed size. The code table is made of 39 patterns of length greater than or equal to 2 (cf. Fig. 3) and 465 items (singletons). Singletons do not compress (but are needed for lossless representation). The relative compressed size (length of the final code table divided by the length of the standard code table) is 82.1% $(= 13,179/16,049)$[6]. In comparison, in the study of Vreeken et al. [27], only 4 out of 27 datasets have such a poor compression. This can be explained by Arno's relatively unique publications and collaborations throughout his rich career. This is also illustrated by the 356 items in the dataset DATA-AS-DBLPK (including 31 with a frequency greater than 2) that never appear in the 39 patterns of the code table greater than or equal to 2.

However, close collaboration with Jilles and Matthijs, his two most frequent co-authors, changes the picture from 2006 onwards. This is reflected in the over-representation in the code table of patterns with his two favorite co-authors and keywords around MDL. A natural follow-up question is therefore to investigate this relationship: do Jilles and Matthijs follow Arno's scientific path or not? Answering this question is the goal of the next section.

4.3   Contrasting Arno's Publications to Those of Jilles and Matthijs: Abstract Contents

To answer RQ3, in this section, we additionally explore data from Matthjs van Leeuwen and Jilles Vreeken, former PhD students of Arno Siebes. Our goal is to determine whether Matthijs and Jilles stay in the lane that they explored under their supervisor, or branched out into new areas. We focus on the abstracts of papers since those offer richer semantic information for each paper than only the keywords. Therefore we use DATA-AS-ABSTRACT, DATA-JV-ABSTRACT and DATA-MVL-ABSTRACT (cf. Section 3), from which we remove the abstracts of papers all three co-authored.

Comparing Closed Sets Mined from Paper Abstracts. Using LCM 5.3[7], we mine all frequent closed itemsets of at least size two occurring at least twice in Arno's, Matthijs' and Jilles' paper abstract data. This results in 1169 itemsets for Arno, 16422 for Jilles, and 3480 for Matthijs, in a rather nice illustration of

---

[6] With SLIM on the datasets DATA-JV-DBLPK and DATA-MVL-DBLPK, we compute that Jilles and Matthijs have relative compressed sizes comparable to those of Arno, with 87.6% $(= 23,017/26,276)$ and 86.1% $(= 13,882/16,121)$ respectively. However, their code tables contain far more patterns having at least two items, with 67 for Jilles and 48 for Matthijs.

[7] https://research.nii.ac.jp/~uno/codes.htm.

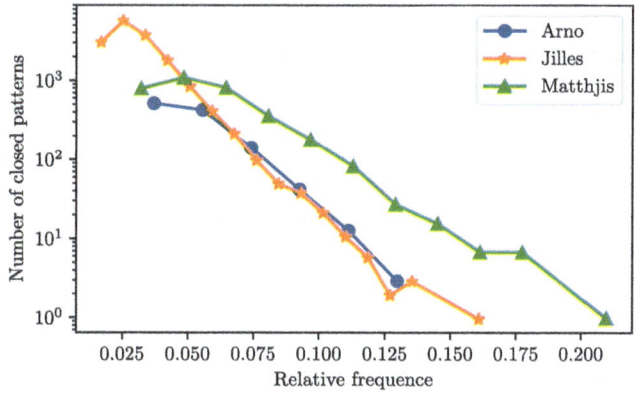

Fig. 4. Support distributions for the paper abstract datasets

how mining individual local patterns becomes overwhelming, *even if* one uses a condensed representation.

As a first, quantitative, point we invite you to have a look at Fig. 4. Starting from the result sets of the three mining operations, we sort the patterns into bins according to their absolute support (starting from 2), calculate relative support to have comparable values[8], and plot the distribution. Relative support values are indicated in the X-axis. The Y-axis shows the number of closed patterns for a given support value (note the logarithmic scale).

As we can see, Matthijs is more consistent in the co-occurring words he uses, with his graph shifted to the right w.r.t. Arno and Jilles. We also see a much larger number of word combinations with low support for Jilles, which can be explained by the fact that he is author of quite a few more publications than Arno and Matthijs (cf. Table 1).

But what we are actually interested in are the bags-of-words that are shared between Arno and his former PhD students, as well as the bags-of-words (denoted *bow*) that are unique to each of them.

Regarding the first, consider Table 3a. It lists frequent closed bags-of-words shared by abstracts of all three authors—Arno, Jilles, Matthijs—sorted on how high their support was among Arno's result (we will perform the same analysis for Jilles and Matthijs). Notably, while the highest-ranked bow (with rank 0) in Arno's abstracts occurs in the shared bows, this is neither the case for Jilles nor for Matthijs. In other words, their highest-ranked bows are unique to them. It shows a number of interesting phenomena: (1) mining, set is the most frequent bow in Arno's paper abstracts[9] and while it continues to play an important role in Matthijs' abstracts, it drops off quite a bit for Jilles; (2) data, mining is very highly ranked for both Arno and Matthijs but somewhat less so for Jilles; (3) the bow with the highest average rank is mining, pattern, important to all

---

[8] Given the different sizes of the underlying datasets.

[9] Given that we are computer scientists, we start counting at 0!.

three authors; (4) two other bows that are important to all three authors are data, set and pattern, set; (5) Arno and Jilles often seem to discuss at least two topics in their work, given that as, well occurs often for both, whereas Matthijs seems to have written single-issue papers; (6) the biggest difference in focus between Arno's works and those of his PhD students seems to be that he cared about providing models and offer approaches that allow certain forms of mining.

It would be unfair to consider these results only from the point-of-view of the promoter, however, which is why we repeat this exercise in Table 3b and 3c, as promised.

As we can see, Jilles really likes mining, patterns as, well as data and sets. A notable difference is that his bows involve verbs that do not show up in Arno's abstracts: propose, use, find. In Matthijs' bows, we see the first three-term set—data, mining, set—as well as the first negation—not, such.

As mentioned above, Arno's highest-ranked bow occurs in the shared bows, which is not true for Jilles and Matthijs. We therefore add a final table (Table 4) showing the highest-ranked unique bows for each author. This last table is arguably the most interesting one: (1) Arno, Jilles, and Matthijs worked together on using the MDL principle to select subsets of patterns (most famously via KRIMP [26]). Even though the removal of co-authored papers from the data removes those as possible sources for "mdl", their PhD work echoes in Jilles' and Matthijs' later work, however, either explicitly, e.g. in description, mdl, principle, or implicitly: length,minimum. (2) Jilles' abstracts tend to repeat certain terms that are almost required to get published in highly ranked venues nowadays: art, state, efficient(ly), extensive. (3) Matthijs surprisingly is the only one who often mentions explicit tasks: subgroup discovery and anomaly detection. (4) And in Arno's abstracts we find terms (and combinations) that hark back to an earlier time in data mining, and have fallen out of use: inductive model, constraint, property, even rule. Terms that roamed wide and free before the deep learning "revolution".

Using KRIMP to Find Compressing Bows. But everything we have done in this section so far is arguably representative of the knowledge discovery process as envisioned at the dawn of data mining: mining local patterns, ranking them according to some criterion, sifting through them manually, and trying to understand what they tell us. It would be much better to use a method that actually relates patterns to each other, and for this task, we turn to KRIMP, the algorithm introduced in the first paper co-authored by Arno Siebes, Jilles Vreeken, and Matthijs van Leeuwen: "Itemsets that compress" [26]. Doing so should allow us to cut down on the redundancy we have observed in Table 3a to 3c, and bring additional bows to our attention.

Working with the same dataset, we therefore extracted the bows that *compress* the respective paper abstract sets. We had to increase the support threshold to four, however, since we otherwise retrieve very long bags-of-words that marvelously compress abstracts representing a conference publication (or two very closely related ones) and the journal publication that eventually resulted from it. We show the results in Table 5 in Appendix B.

Table 3. Top-20 shared paper abstract bows

(a) Arno's perspective				(b) Jilles' perspective				(c) Matthijs' perspective			
Bow	$Rank_A$	$Rank_J$	$Rank_M$	Bow	$Rank_A$	$Rank_J$	$Rank_M$	Bow	$Rank_A$	$Rank_J$	$Rank_M$
mining,set	0	470	10	mining,pattern	13	1	2	mining,pattern	13	1	2
approach,use	2	68	27	as,well	14	3	2511	data,mining	3	46	3
data,mining	3	46	3	data,propose	154	10	36	find,set	104	469	4
mining,pattern	13	1	2	algorithm,data	65	14	201	algorithm,approach	214	67	7
as,well	14	3	2511	pattern,set	43	25	33	data,set	18	26	9
mining,model	16	881	62	data,set	18	26	9	mining,set	0	470	10
approach,mining	17	1976	333	problem,set	137	27	68	model,propose	155	259	11
data,set	18	26	9	find,such	106	34	171	approach,use	2	68	27
model,set	19	471	145	data,mining	3	46	3	pattern,set	43	25	33
approach,set	20	1980	61	data,model	59	47	59	data,mining,set	204	1756	35
different,model	30	2840	214	algorithm,approach	214	67	7	data,propose	154	10	36
model,term	41	614	1318	approach,use	2	68	27	approach,find	102	1028	42
pattern,set	43	25	33	model,use	66	87	105	find,introduce	107	2530	43
mining,problem	49	458	334	data,present	213	132	631	data,model	59	47	59
data,model	59	47	59	set,such	74	148	169	approach,set	20	1980	61
algorithm,data	65	14	201	model,propose	155	259	11	mining,model	16	881	62
model,use	66	87	105	data,propose,set	398	260	152	problem,set	137	27	68
model,such	72	909	168	data,information	298	342	1379	find,mining	103	1748	73
set,such	74	148	169	mining,problem	49	458	334	not,such	682	4578	100
allow,mining	93	2813	1106	find,set	104	469	4	model,use	66	87	105

Visually, the most striking result is how much shorter Arno's bows are compared to Matthijs' and (in particular) Jilles'. This is also reflected in the compression rate of the three datasets: Arno's abstract set can be compressed to 71%, Jilles' to 63.17%, and Matthijs' to 63.82%. This means that there is less redundancy to be found in Arno's abstracts but without a more detailed analysis of the underlying papers, we can only speculate as to the reasons: Jilles and Matthijs might publish more related conference papers than Arno did and does, or they might publish for a longer period on a particular topic than Arno did (and had to do).

The KRIMP results tease out the importance of Kolmogorov complexity[10], minimum description length principle, and mdl for Jilles and Matthijs, whereas neither one appears in Arno's top-20. As seen before, we find the indicator terms for more recent publications—state-of-the-art, synthetic data, high dimensional, extensive etc. for the younger researchers.

The reduced redundancy allows some new topics to emerge, such as that Arno apparently worked on time series (similarity) at some point[11], or Jilles' interest in high-dimensional data.

All in all, using KRIMP recovers arguably some knowledge we have already identified comparing closed bows but with less redundancy, and richer patterns.

Edited Publications. Edited publications (cf. Table 1) seem to reveal another difference between Arno on the one hand and Jilles and Matthijs on the other.

---

[10] The theoretical underpinnings of MDL.

[11] Two papers, from 1996, 1997, describing the DEGAS model are the source of the by far longest compressing bow if we leave the minimum support threshold at two! :).

Table 4. Top-25 unique paper abstract bows for each author

Arno	Jilles	Matthijs
database,model	art,state	approach,propose
approach,present	experiment,extensive	detection,method
base,use	data,pattern	datasets,method,result
mining,research	discover,well	approach,datasets
approach,order	algorithm,use	demonstrate,propose
base,model	length,minimum	model,novel
base,set	pattern,such	algorithm,approach,propose
base,present	extensive,problem	description,mdl,principle
algorithm,database	discover,extensive	approach,propose,result
mining,use	discover,efficiently	description,find
algorithm,different	art,problem	approach,datasets,propose
paper,present	problem,term	discovery,subgroup
define,type	describe,pattern	approach,demonstrate,propose
behaviour,rule	x,y	complex,result
mining,order	set,well	description,mdl,model,principle
information,present	algorithm,problem	anomaly,detection,method
combine,very	data,experiment	approach,model,propose
set,show	experiment,problem	approach,find,propose
behaviour,constraint	efficient,propose	find,introduce,set
property,set	extensive,set	description,introduce,model
inductive,model	extensive,propose	approach,description,introduce
novel,term	efficiently,experiment	datasets,method,model
allow,suitable	efficiently,extensive	data,mining,search
decomposition,similarity	art,extensive,state	discovery,search
approach,mining,set	real-world,synthetic	approach,datasets,demonstrate

The number of edited publications, normalized on a same period of time, is about twice for Arno as for Jilles and Matthijs. However, these numbers are small and this trend is not strongly supported by the data. Interestingly, using edited publications, SLIM highlights several collaborations, including one with Katharina Morik for two Dagstuhl Seminar Proceedings. It also brings out three authors over the period 2004–2010 in the code table absent when using the original dataset DATA-AS-DBLPK: José Maria Peña Sánchez and Joost N. Kok (for two entries) and Jean-François Boulicaut (for four entries).

5   What Have We Learned?

In this chapter, we have explored Arno-centric bibliographic data in order to highlight some trends in his scientific career. We based our analysis on pattern mining methods, expecting them to allow us to derive interpretable insights into the data. In general terms, we are glad to report that our expectations have been met, and we could characterize the publication of the person being lauded in this collection, i.e. Arno.

The uncovered patterns show us that this particular leopard did indeed change his spots (several times). Around 1999, we see a first turning point with the keywords KDD and knowledge discovery (recalling that Arno was a pioneer, presenting himself as a data mining researcher as early as 1996 [23]). The keywords frequent itemsets, association rules and data mining occur in the period 1999–2003. A wave of keywords related to pattern mining start appearing in the period 2004–2010. This period (and even beyond) is strongly marked by work based on the MDL principle. Arno's close collaboration with Jilles and Matthijs, his two most frequent co-authors, changes the picture from 2006 onwards. ICDM and ECML-PKDD are two major venues for Arno.

Arno also clearly has instilled the importance of blazing their own trail and not following too closely in their mentor's foot steps into Jilles and Matthijs. Both in terms of frequent bags-of-words derived from their paper abstracts and in terms of compressing bags-of-words selected using KRIMP, there is a noticeable difference between Arno and his former PhD students (and between Jilles and Matthijs themselves). Related to the latter point, Arno's data do not compress well, worse than those of Jilles and Matthijs. On the one hand, this is the sign of a researcher with eclectic research interests, leading to a particularly varied publication activity, and collaborations throughout his rich career. On the other hand, the academic environment, the pressure to publish (often and much), and freedom to explore many different topics have changed over the decades so that some of the differences may not be due to choice by the younger researchers.

This is also where we need to temper the general statement from above a bit: while pattern mining methods clearly continue to have their place in the data analysis tool box, the pipeline is not as easy as Data → Pattern Mining → Profit! Instead, interpretable results *do* require a lot of interpretation, knowledge of the underlying data, and the problem setting at hand, to arrive at meaningful information.

As indicated above, the code for the data collection and preparation is freely accessible. This means that we can do this exercise again twenty years from now when Jilles and Matthijs retire.[12]

Acknowledgments. The authors thank Djawad Bekkoucha for his help with data collection.

---

[12] Or maybe a bit earlier when Arno Knobbe's time has come.

A   DBLP Data

```
PREFIX rdfs: <http://www.w3.org/2000/01/rdf-schema#>
PREFIX dblp: <https://dblp.org/rdf/schema#>

select ?paper ?title ?year (COALESCE(?publishedIn, "") as ?publication)
(group_concat(?author) as ?authors) (group_concat(?label) as ?author_labels)
where {
    ?paper dblp:title ?title .
    ?paper dblp:yearOfPublication ?year .
    ?paper dblp:authoredBy ?author .
    ?author rdfs:label ?label .
    optional {
        ?paper dblp:publishedIn ?publishedIn .
    }
    .
    {
        select distinct ?paper where {
            ?paper dblp:authoredBy ?author .
            {
            select distinct ?author where {
                ?paper dblp:authoredBy <https://dblp.org/pid/s/ArnoSiebes> .
                ?paper dblp:authoredBy ?author
                }
            }
        }
    }
}
group by ?paper ?title ?year ?publishedIn
```

Listing 3.1. SPARQL request to query DBLP

B   Compressing Bows Mined with Krimp

Table 5. Top-20 most-compressing bows for paper abstracts of our three authors

Arno	Jilles	Matthijs
mining, set, pattern, major	search, function, optimization, heuristic, reliable, algorithmic, high-dimensional, bounding	method, result, detection, system, anomaly, perform, prediction, collection, step, event, widely, log
find, similarity, series, decomposition	base, causal, complexity, infer, kolmogorov, observational, margin	method, result, datasets, detection, anomaly, outperform, accuracy, detect
data, mining, use	also, learn, capture, form, robustness, communication, play	model, find, description, principle, length, minimum, mdl
model, database, inductive	algorithm, identify, most, present, itemsets, interesting	method, detection, aim, state-of-the-art, anomaly, accuracy
model, different, introduce	use, function, evaluation, case, variable, wide	method, result, approach, propose, datasets
set, base, design	method, work, include, state-of-the-art, reason, promise	set, attribute, most, first, same
such, similarity, decomposition	problem, extensive, term, best, optimization	model, novel, provide, first, dub
time, series, representation	pattern, mine, event, sequential, occurrence	approach, propose, novel, algorithm, real-world
approach, use	data, model, algorithm, synthetic, principle	find, other, only, present, good
provide, algorithm	data, mining, many, analysis, multiple	description, principle, length, minimum
approach, order	pattern, synthetic, consider, meaningful, space	data, pattern, mining, discover
well, as	model, mdl, formalize, greedy, lossless	description, principle, mdl, compression
many, mine	graph, network, present, make, highlight	approach, propose, novel, most
model, inductive	base, complexity, type, kolmogorov, instantiation	data, evaluate, analysis, infer
present, paper	description, length, minimum, principle	model, find, subgroup, target
behaviour, rule	problem, extensive, art, state	approach, search, algorithm, exist
define, type	problem, algorithm, synthetic, propose	search, discovery, algorithm, space
allow, suitable	problem, practice, art, introduce	learn, high, system, experiments
term, novel	find, interpretable, succinct, lead	approach, propose, algorithm
combine, very	synthetic, only, include, other	data, demonstrate, only

C   Redescriptions

Table 6. Redescriptions mined from Arno's publication data

q_L	q_R	J	|supp|	pV
ArnoJKnobbe and inductive-logic-programming				
R0 ArnoJKnobbe	inductive-logic-programming	1.000	4	0.000
R9 ArnoJKnobbe ∧ PKDD	inductive-logic-programming	1.000	4	0.000
R10 ArnoJKnobbe	PKDD ∧ inductive-logic-programming	1.000	4	0.000
R17 ArnoJKnobbe	inductive-logic-programming∧ 1999–2003	1.000	4	0.000
R11 PKDD ∧ inductive-logic-programming	ArnoJKnobbe ∧ 1999–2003	1.000	4	0.000
JillesVreeken				
R12 JillesVreeken ∧ RoelBertens	synthetic-data	1.000	4	0.000
R18 JillesVreeken ∧ synthetic-data	RoelBertens	0.800	4	0.000
R22 JillesVreeken	mdl	0.250	4	0.024
2004–2010				
R21 mdl ∨ transaction-database	2004–2010	0.333	10	0.005
R25 JillesVreeken	2004–2010	0.235	8	0.080
R26 RonnieBathoorn ∨ ICDM	2004–2010	0.276	8	0.006
R27 ICDM	2004–2010	0.138	4	0.050
R28 transaction-database	2004–2010	0.138	4	0.050
R6 MatthijsvanLeeuwen ∨ RonnieBathoorn	2004–2010	0.400	12	0.002
2011–2025				
R13 RoelBertens ∨ pattern-mining	2011–2025	0.435	10	0.001
R20 pattern-mining ∨ synthetic-data	2011–2025	0.391	9	0.001
R24 synthetic-data	2011–2025	0.182	4	0.021

Table 7. Redescriptions mined from Matthijs' data

	$q_\mathbf{L}$	$q_\mathbf{R}$	J	|supp|	pV
	ZhongLi, anomaly-detection				
R0	ZhongLi0002	anomaly-detection	0.727	8	0.000
R5	anomaly-detection	event-log	0.500	4	0.000
R12	ZhongLi0002	event-log	0.364	4	0.001
R18	ZhongLi0002	CoRR	0.214	6	0.051
	Arno, Jilles				
R1	SandervanRijn	SebastianSchmitt	0.667	4	0.000
R9	SandervanRijn	SebastianSchmitt ∧ ThomasBck	0.667	4	0.000
R13	SebastianSchmitt	SandervanRijn ∧ ThomasBck	0.667	4	0.000
	Arno, Jilles				
R2	ArnoSiebes	2004–2010	0.667	8	0.000
R8	ArnoSiebes	JillesVreeken	0.636	7	0.000
R11	JillesVreeken	2004–2010	0.429	6	0.001
	LucDeRaedt, SiegfriedNijssen, ThanhLeVan				
R3	SiegfriedNijssen	ThanhLeVan	0.625	5	0.000
R4	LucDeRaedt ∧ ThanhLeVan	SiegfriedNijssen	0.625	5	0.000
R10	LucDeRaedt∧SiegfriedNijssen	ThanhLeVan	0.714	5	0.000
R17	LucDeRaedt	pattern-mining	0.278	5	0.013
R19	LucDeRaedt	VladimirDzyuba	0.267	4	0.014
R20	LucDeRaedt ∨ minimum-description-length	data-mining	0.333	8	0.005

Table 8. Redescriptions mined from Jilles' data

q_L	q_R	J	|supp|	pV
ZhongLi, anomaly-detection				
R0 ZhongLi0002	anomaly-detection	0.727	8	0.000
R5 anomaly-detection	event-log	0.500	4	0.000
R12 ZhongLi0002	event-log	0.364	4	0.001
R18 ZhongLi0002	CoRR	0.214	6	0.051
Arno, Jilles				
R1 SandervanRijn	SebastianSchmitt	0.667	4	0.000
R9 SandervanRijn	SebastianSchmitt $\wedge$ ThomasBck	0.667	4	0.000
R13 SebastianSchmitt	SandervanRijn $\wedge$ ThomasBck	0.667	4	0.000
Arno, Jilles				
R2 ArnoSiebes	2004–2010	0.667	8	0.000
R8 ArnoSiebes	JillesVreeken	0.636	7	0.000
R11 JillesVreeken	2004–2010	0.429	6	0.001
LucDeRaedt, SiegfriedNijssen, ThanhLeVan				
R3 SiegfriedNijssen	ThanhLeVan	0.625	5	0.000
R4 LucDeRaedt $\wedge$ ThanhLeVan	SiegfriedNijssen	0.625	5	0.000
R10 LucDeRaedt $\wedge$ SiegfriedNijssen	ThanhLeVan	0.714	5	0.000
R17 LucDeRaedt	pattern-mining	0.278	5	0.013
R19 LucDeRaedt	VladimirDzyuba	0.267	4	0.014
R20 LucDeRaedt $\vee$ minimum-description-length	data-mining	0.333	8	0.005

References

1. Agrawal, R., Imielinski, T., Swami, A.N.: Mining association rules between sets of items in large databases. In: SIGMOD, pp. 207–216 (1993)
2. Berthold, M.R., Morik, K., Siebes, A.: 07181 abstracts collection – parallel universes and local patterns. In: Berthold, M.R., Morik, K., Siebes, A. (eds.) Parallel Universes and Local Patterns. Dagstuhl Seminar Proceedings (DagSemProc), vol. 7181, pp. 1–15. Schloss Dagstuhl - Leibniz-Zentrum für Informatik, Dagstuhl, Germany (2007)
3. Boulicaut, J.-F., Bykowski, A., Rigotti, C.: Free-sets: a condensed representation of Boolean data for the approximation of frequency queries. Data Min. Knowl. Disc. **7**(1), 5–22 (2003)
4. Boulicaut, J.-F., Morik, K., Siebes, A.: 04161 abstracts collection – detecting local patterns. In: Boulicaut, J.-F., Morik, K., Siebes, A. (eds.) Detecting Local Patterns. Dagstuhl Seminar Proceedings (DagSemProc), vol. 4161, pp. 1–8. Schloss Dagstuhl - Leibniz-Zentrum für Informatik, Dagstuhl, Germany (2006)

5. Buzmakov, A.: Formal concept analysis and pattern structures for mining struc-
   tured data. PhD thesis, Universite de Lorraine (2015)
6. Calders, T., Goethals, B.: Mining all non-derivable frequent itemsets. In: Elomaa,
   T., Mannila, H., Toivonen, H. (eds.) PKDD 2002. LNCS, vol. 2431, pp. 74–86.
   Springer, Heidelberg (2002). https://doi.org/10.1007/3-540-45681-3_7
7. Dao, T., Kuo, C., Ravi, S.S., Vrain, C., Davidson, I.: Descriptive clustering: ILP and
   CP formulations with applications. In: Lang, J., (ed.) Proceedings of the Twenty-
   Seventh International Joint Conference on Artificial Intelligence, IJCAI 2018, July
   13-19, 2018, Stockholm, Sweden, pp. 1263–1269 (2018)
8. De Raedt, L., Zimmermann, A.: Constraint-based pattern set mining. In: SDM
   (2007)
9. de Solla Price, D.J., Beaver, D.: Collaboration in an invisible college. Am. Psychol.
   21(11), 1011 (1966)
10. Durand, N., Crémilleux, B.: ECCLAT: a new approach of clusters discovery in
    categorical data. In: 22nd International Conference on Knowledge Based Systems
    and Applied Artificial Intelligence (ES'02), pp. 177–190, Cambridge (2002)
11. Galbrun, E., Miettinen, P.: Redescription Mining. Springer, Cham (2017). https://
    doi.org/10.1007/978-3-319-72889-6_3
12. Grünwald, P.: Minimum description length tutorial. Adv. Minimum Description
    Length: Theory Appl. **5**, 1–80 (2005)
13. Kifer, D., Gehrke, J., Bucila, C., White, W.M.: How to quickly find a witness. In:
    PODS, pp. 272–283 (2003)
14. Knobbe, A., Crémilleux, B., Fürnkranz, J., Scholz, M.: From local patterns to
    global models: the lego approach to data mining. In: Fürnkranz, J., Knobbe, A.,
    (eds.) From Local Patterns to Global Models: Proceedings of the ECML/PKDD-08
    Workshop, pp. 1–16 (2008)
15. Knobbe, A.J., Ho, E.K.Y.: Pattern teams. In: PKDD, pp. 577–584 (2006)
16. Kostoff, R.N., del Rio, J.A., Humenik, J.A., Garcia, E.O., Ramirez, A.M.: Citation
    mining: integrating text mining and bibliometrics for research user profiling. J. Am.
    Soc. Inform. Sci. Technol. **52**(13), 1148–1156 (2001)
17. Kuznetsov, S.O.: Galois connections in data analysis: contributions from the soviet
    era and modern Russian research. In: Ganter, B., Stumme, G., Wille, R. (eds.) For-
    mal Concept Analysis. LNCS (LNAI), vol. 3626, pp. 196–225. Springer, Heidelberg
    (2005). https://doi.org/10.1007/11528784_11
18. Ley, M.: The DBLP computer science bibliography: evolution, research issues, per-
    spectives. In: Laender, A.H.F., Oliveira, A.L. (eds.) SPIRE 2002. LNCS, vol. 2476,
    pp. 1–10. Springer, Heidelberg (2002). https://doi.org/10.1007/3-540-45735-6_1
19. Pasquier, N., Bastide, Y., Taouil, R., Lakhal, L.: Discovering frequent closed item-
    sets for association rules. In: Beeri, C., Buneman, P. (eds.) ICDT 1999. LNCS,
    vol. 1540, pp. 398–416. Springer, Heidelberg (1999). https://doi.org/10.1007/3-
    540-49257-7_25
20. Pei, J., Han, J., Lakshmanan, L.V.S.: Pushing convertible constraints in frequent
    itemset mining. Data Min. Knowl. Discov. **8**(3), 227–252 (2004)
21. Rissanen, J.: Modeling by shortest data description. Automatica **14**(5), 465–471
    (1978)
22. Siebes, A.: Queries for data analysis. In: Hollmén, J., Klawonn, F., Tucker, A.
    (eds.) IDA 2012. LNCS, vol. 7619, pp. 7–22. Springer, Heidelberg (2012). https://
    doi.org/10.1007/978-3-642-34156-4_3
23. Siebes, A., Tsur, S., Ullman, J., Vieille, L., Zaniolo, C.: Deductive databases: chal-
    lenges, opportunities and future directions. In: Pedreschi, D., Zaniolo, C. (eds.)

LID 1996. LNCS, vol. 1154, pp. 223–229. Springer, Heidelberg (1996). https://doi. org/10.1007/BFb0031743

24. Smets, K., Vreeken, J.: Slim: directly mining descriptive patterns. In: Proceedings of the Twelfth SIAM International Conference on Data Mining, Anaheim, California, USA, April 26-28, 2012, pp. 236–247 (2012)
25. Soulet, A., Crémilleux, B.: Adequate condensed representations of patterns. Data Min. Knowl. Disc. **17**(1), 94–110 (2008)
26. van Leeuwen, M., Vreeken, J., Siebes, A.: Compression picks item sets that matter. In: PKDD, pp. 585–592 (2006)
27. Vreeken, J., Van Leeuwen, M., Siebes, A.: KRIMP: mining itemsets that compress. Data Min. Knowl. Disc. **23**, 169–214 (2011)
28. Whalen, J.M., Nelson, D.J., Whalen, R.J., Provencher, M.T.: Coaching, mentorship, and leadership lessons learned from professional football. Clin. Sports Med. **42**(2), 291–299 (2023)
29. Yan, X., Han, J.: GSPAN: graph-based substructure pattern mining. In: ICDM, pp. 721–724. Japan (2002)
30. Zaki, M.J.: Sequence mining in categorical domains: incorporating constraints. In: Proceedings of the Ninth International Conference on Information and Knowledge Management, pp 422–429 (2000)
31. Zaki, M.J.: Efficiently mining frequent trees in a forest. In: KDD, pp. 71–80 (2002)
32. Zaki, M.J., Hsiao, C.-J.: CHARM: an efficient algorithm for closed itemset mining. In: 2nd SIAM International Conference on Data Mining (2002)
33. Zaki, M.J., Ramakrishnan, N.: Reasoning about sets using redescription mining. In: Proceedings of the Eleventh ACM SIGKDD International Conference on Knowledge Discovery in Data Mining, pp. 364–373 (2005)

Snor: Simpler Descriptions Through Overlapping Patterns

Matthijs van Leeuwen[1(✉)] and Jilles Vreeken[2]

[1] Universiteit Leiden, Leiden, The Netherlands
m.van.leeuwen@liacs.leidenuniv.nl
[2] CISPA Helmholtz Center for Information Security, Saarbrücken, Germany
jv@cispa.de

Abstract. The pattern explosion is a well-known problem that refers to the humongous numbers of results discovered by traditional pattern mining algorithms. One line of research that combats this, pioneered by KRIMP, is to employ the minimum description length (MDL) principle for selecting small sets of characteristic patterns.

When KRIMP was proposed twenty years ago, it was shown to work well but also had several limitations. Over time its major shortcomings have been addressed, except for one: the cover function still only considers non-overlapping patterns to describe transactions. While efficient, this may result in redundant pattern sets, as we need *combinations of patterns* rather than just the *true patterns* to succinctly describe the data.

In this paper, we propose SNOR, a cover function that enables simpler descriptions by allowing overlapping patterns in the cover of a transaction. Our key observation is that the cover problem is an instance of (weighted) set cover, and we can hence use a greedy method to approximate it. SNOR can be straightforwardly combined with existing search algorithms for finding itemset code tables, such as KRIMP and SLIM.

In experiments on 18 benchmark datasets, we demonstrate that SNOR leads to simpler descriptions, i.e., allowing overlap leads to smaller code tables while compression stays the same. Further, classification accuracy and computation time remain almost the same.

1 Introduction

Exactly 20 years ago, in September 2005, Prof. Dr. Arno Siebes coined an idea to the authors of this paper, who had then just started as PhD students under his supervision. The idea was simple: let's use the Minimum Description Length (MDL) principle [12,27] to defeat the pattern explosion in frequent itemset mining [1]. After sorting out a few further details, this resulted in KRIMP– originally published at SIAM SDM 2006 [30] as a method and algorithm without a name, and later extended and given its name in 2011 [36].

KRIMP[1] was the first approach to combine the MDL principle and pattern mining. It introduced the notion of a 'code table', a model containing pairs of

---

[1] KRIMP means 'to shrink' in Dutch.

© The Author(s), under exclusive license to Springer Nature Switzerland AG 2026
M. van Leeuwen and J. Vreeken (Eds.): Arno Siebes Festschrift, LNCS 16067, pp. 56–74, 2026.
https://doi.org/10.1007/978-3-032-03028-3_4

patterns and code words that can be used to losslessly compress data. Compressing data was never the goal but rather a means to an end: following the MDL principle, the slogan of the approach became that 'the best set of patterns is the one that compresses the data best'. As such, KRIMP introduced a principled and rigorous approach for model selection in (unsupervised) data mining tasks such as data characterisation using frequent patterns.

A series of papers on related but data mining different tasks [21,22,31] followed, and the approach was also picked up by others (see Sect. 2). Over time, MDL-based pattern mining became a successful research area on itself, as can be witnessed from, for example, the 2022 survey dedicated to the topic by Galbrun [11], and the mention of the research area as a 'flourishing application' of MDL in the 2019 'MDL revisited' paper by Grünwald and Roos [13].

As the first method for MDL-based pattern mining, KRIMP had clear limitations. First, although the MDL principle prescribes using a lossless encoding, the initial proposal [30] implicitly, and the later version explicitly [36], swept some bits under the rug. Second, the filtering approach employed by the KRIMP search heuristic had an obvious downside in needing frequent itemsets as candidates. Not only does this require choosing a minimum support threshold, but also incurs long runtimes thanks to the same pattern explosion that KRIMP aims to solve. This problem was addressed by SLIM[2] [32], which generates candidates on-the-fly by considering how patterns currently in the code table are being used.

Third, the explicit inclusion of code words in the code table was suboptimal; not only did this lead to the first-mentioned problem, but in fact turned out to be unnecessary. That is, in 2015, Budhathoki and Vreeken [6] showed how we can use prequential codes to (near) optimally encode data without having to first explicitly transmit the codes in the code table. Importantly, they showed that using their scheme the total encoded length can still be computed efficiently, maintaining efficient code table search with the improved encoding.

Fourth, KRIMP disallows overlap between patterns that describe a transaction; it employs a 'cover' function that iteratively selects the longest and most frequent pattern in the model that does not overlap with previously selected patterns, until the entire transaction is described. This choice was because of efficiency (maintaining a sorted list is fast), because the ideas we tried at the time did not work, and because it was unclear whether overlap makes sense in light of MDL (why describe items multiple times if once suffices?). Ever since, non-overlapping cover functions have been taken for granted, and cover functions for other pattern and data types have followed the original idea. This is a pity, because allowing overlap does result in *smaller* pattern sets.

In this paper, we once and for all address the problem of overlapping covers. We propose an alternative cover function called SNOR[3] that allows for the cover of a transaction to contain overlapping patterns. Our key observation is that the cover problem—i.e., finding a subset of code table elements that together cover a given transaction—can be regarded as an instance of the (weighted) set

---

[2] SLIM means 'smart' in Dutch.
[3] SNOR means 'moustache' in Dutch.

cover problem. Following this observation, we propose to replace the original cover function by the well-known greedy algorithm for approximating the set cover problem, iteratively finding that pattern that covers most of the (currently uncovered part of the) transaction.

The main contributions of this paper are:

(a) We revisit KRIMP 20 years after its introduction, discuss its strengths and limitations, and describe improvements that have been proposed.
(b) We propose SNOR, a greedy approximation to the set cover problem that covers a transaction with overlapping itemsets.
(c) We instantiate KRIMP and SLIM using SNOR and empirically demonstrate that this leads to simpler descriptions. Averaged over 18 benchmark datasets, it leads to smaller code tables that compress and classify as well as before.

We review related work in Sect. 2, and revisit MDL for pattern mining in Sect. 3. We discuss the classic COVER function and introduce SNOR for covering with overlap in Sect. 4. After that, we will briefly describe the KRIMP and SLIM algorithms (Sect. 5), followed by experiments (Sect. 6) and conclusions (Sect. 7).

2   Related Work

In this section we will briefly describe some notable papers that are closely related to KRIMP, either as a direct improvement, by using KRIMP code tables for another task, or by extending KRIMP to another pattern type. For a comprehensive survey on MDL for pattern mining until 2022, see Galbrun [11].

Krimp Improvements. As already mentioned, SLIM [32] and DIFFNORM [6] directly built and improved on KRIMP. SLIM maintained the exact same code table model and coding as was introduced for KRIMP, but vastly improved the search algorithm by eliminating the need for a candidate set of itemsets as input. Instead, it generates candidate patterns on-the-fly by combining pairs of patterns currently in the code table and iteratively adds the generated candidate pattern that improves compression most. In addition, it uses heuristics that estimate compression gain to speed-up the candidate generation and search process.

DIFFNORM [6] uses KRIMP code tables to characterise multiple databases by looking for their 'norm' and 'differences', represented by a set of code tables modelling both sets of databases and individual databases. Moreover, it introduced prequential coding for encoding the data given the code table, replacing the Shannon-Fano coding and resulting in better overall compression by removing the need to explicitly encode code words (or usages) in the code tables.

Widened KRIMP [28] uses the concept of 'widening' to diversify the search for code tables. Instead of using a single greedy hillclimber, the goal is to pursue multiple paths in the search space in parallel, thereby hoping to find a better solution. By employing parallelisation, the goal is to achieve this without increasing runtime compared to the hillclimber.

Tasks Using Krimp and Slim. MDL-optimal models have desirable properties, but when KRIMP was first introduced, it was only shown to find pattern sets that compress the data well. To demonstrate that KRIMP (and later SLIM) code tables are indeed characteristic for the data, they were used for a series of different data mining tasks: classification [21], difference characterisation [35], change detection in data streams [20], clustering [22], structure functions [16], anomaly detection [2,5,31], denoising [17], and providing insights into models [10,14].

In this paper we revisit the classification task, to investigate the effect of allowing overlap within the cover of a transaction on classification accuracy. It would be interesting to also revisit other tasks with our proposed method.

Other Pattern and Data Types. Over the years the MDL for pattern mining approach that started with KRIMP has been extended to numerous other pattern and data types. One direction that has received considerable attention, in several settings, is considering richer pattern languages, such as rules [26,34,39] and general dependencies [15]. Other extensions often consider more complex data and corresponding pattern types, e.g., relational data [18], sequences [4,8,33], process logs [38], graphs [3,19,29], and numerical data [24].

3   MDL for Itemset Mining—Revisited

In this section we introduce notation, the MDL principle, and the problem of finding the MDL-optimal code table for a given transaction database.

3.1   Notation

In this paper we consider transaction databases. Let $\mathcal{I}$ be a set of items, e.g., the products for sale in a shop. A transaction $t \in \mathcal{P}(\mathcal{I})$ is a set of items that, e.g., represent the items a customer bought. A database D over $\mathcal{I}$ is then a bag of transactions, e.g., the different sale transactions on a given day. We say that a transaction $t \in D$ supports an itemset $X \subseteq \mathcal{I}$, iff $X \subseteq t$. The support of X in D is the number of transactions in the database where X occurs.

All logarithms are to base 2, and by convention we say $0 \log 0 = 0$.

3.2   MDL, a Brief Introduction

The Minimum Description Length principle (MDL) [12], like its close cousin MML (Minimum Message Length) [37], is a practical version of Kolmogorov Complexity [23]. All three embrace the slogan *Induction by Compression*. For MDL, this can be roughly described as follows.

Given a set of models $\mathcal{M}$, the best model $M \in \mathcal{M}$ is the one that minimises

$$L(D, M) = L(M) + L(D \mid M) \,,$$

in which $L(M)$ is the length in bits of the description of M, and $L(D \mid M)$ is the length of the description of the data when encoded with model M.

This is called two-part MDL, or *crude* MDL—as opposed to *refined* MDL, where model and data are encoded together [12]. We use two-part MDL because we are specifically interested in the model: the patterns that give the best description. Further, although refined MDL has stronger theoretical foundations, it cannot be computed except for some special cases.

To use MDL, we have to define what our models $\mathcal{M}$ are, how a $M \in \mathcal{M}$ describes a database, and how we encode these in bits. Note, that in MDL we are only concerned with code lengths, not actual code words.

3.3   The MDL-Optimal Code Table

As models we use the exact same itemset-based code tables as those introduced for KRIMP [30, 36]. That is, a code table is a two-column table with itemsets in the left-hand column, and corresponding code words in the right-hand column.[4]

The actual codes are of no importance: their lengths are. To explain how these lengths are computed, we first need to understand the coding algorithm. A transaction t is encoded by a subset of the itemsets in the code table, which is called its *cover*. A cover function takes a transaction (and code table) as input, and outputs the corresponding cover. We discuss how to instantiate a cover function in Sect. 4. For now, it suffices that a cover $C \subseteq CT$ of a transaction t permits its lossless reconstruction, i.e., $\bigcup_{X \in C} X = t$. Each code table is required to contain at least all singletons, i.e., all single items, to ensure that a cover can be constructed for each (possible) transaction over $\mathcal{I}$. A database is encoded by encoding each transaction $t \in D$ independently.

The *usage* of an itemset $X \in CT$ is the number of transactions $t \in D$ that have X in their cover. The relative usage of $X \in CT$ is the probability that X is used in the encoding of an arbitrary $t \in D$. For optimal compression of D, the higher $\Pr(X \mid D)$, the shorter its code should be. Shannon entropy [7] gives us the length of the optimal prefix code for X as

$$L(X \mid D) = -\log \Pr(X \mid D) ,$$

where

$$\Pr(X \mid D) = \frac{usage(X)}{\sum\limits_{Y \in CT} usage(Y)} .$$

The length of the encoding of transaction is now simply the sum of the code lengths of the itemsets in its cover,

$$L(t \mid CT) = \sum_{X \in cover(t)} L(X \mid CT) .$$

---

[4] Note that we are not using prequential coding for the data, as introduced by DIFFNORM [6]. This is a deliberate choice, as we aim to compare KRIMP and SLIM with and without SNOR. It will make, as the adage goes, for engaging future research to explore how overlap works with more advanced coding schemes.

The size of the encoded database is then the sum of the sizes of the encoded transactions,

$$L(D \mid CT) = \sum_{t \in D} L(t \mid CT) \,.$$

To find the optimal code table, we need to take both the compressed size of the database, and the size of the code table into account. For the size of the code table, we only consider those itemsets that have a non-zero usage. The size of the right-hand column is simply the sum of all the different code lengths. For the size of the left-hand column, note that the simplest valid code table consists only of the single items, to which we refer as the standard code table ST. We encode the itemsets in the left-hand column using the codes of ST.

The encoded size of the code table is then given by

$$L(CT \mid D) = \sum_{\substack{X \in CT \\ usage(X) \neq 0}} L(X \mid ST) + L(X \mid CT) \,.$$

We define the MDL-optimal code table as the one that minimises the total encoded size

$$L(D, CT) = L(CT \mid D) + L(D \mid CT) \,.$$

More formally, we define the problem as follows.

Minimal Coding Set Problem [30] *Let $\mathcal{I}$ be a set of items and let D be a dataset over $\mathcal{I}$. Find the smallest set of patterns $\mathcal{S} \subseteq \mathcal{P}(\mathcal{I})$ such that for the corresponding code table CT the total compressed size, $L(CT, D)$, is minimal.*

The search space of all possible code tables is doubly exponential in the number of unique items. It does not exhibit structure we can use to efficiently find the optimal code table [36]. Hence, we resort to heuristics.

4   Cover Functions

Having defined code table models, code length computations, and the minimal coding set problem, the only missing link is the cover function. We first describe the original cover function, without overlap, and then introduce the new cover function, with overlap.

The idea of a cover function is that we select a set $\mathcal{C} \subseteq CT$ such that the union of the patterns $X \in \mathcal{C}$ form the transaction t at hand, i.e., $\bigcup_{X \in \mathcal{C}} X = t$. The goal of MDL is to describe the data as succinctly as possible, and hence we should choose that $\mathcal{C}$ that minimises the encoded length. If we interpret the code lengths $L(X)$ of sets X as the weights, the problem is hence an instance of Weighted Set Cover, which is known to be NP-hard. In our case, however, things are worse, as the weights depend on how often these codes are used [9]. Rather than exhaustively considering the exponentially many covers of the entire data, or trying to converge to a good solution through expectation-maximization, we resort to heuristics for finding good covers for a transaction t.

4.1 Covering Without Overlap

The standard cover function, denoted COVER, was introduced for KRIMP [30,36] and subsequently used by SLIM [32], DIFFNORM [6], and others. It finds a non-overlapping cover for a given transaction and code table. The idea it pursues is to use as few (and hence long) and as short (and hence often-used) codes as possible to describe t.

To this end, COVER starts with an empty cover, i.e., $\mathcal{C} = \emptyset$, and iteratively adds the first $X \in CT$ such that $X \subseteq t$ and $X \cap C = \emptyset$, where $C = \bigcup \mathcal{C}$. It does so by considering itemsets $X \in CT$ in **Standard Cover Order**, defined as first preferring X of higher cardinality, then those with higher support, and finally, by lexicographical ordering (i.e., using alphabet $\mathcal{I}$, so depending on the original data). By doing so it prefers using long itemsets (such that we need few codes) with high support (such that the chance of these having a short code word is high). By optimising the total encoded size, MDL weeds out candidates that break this idea.

This straightforward cover function is easy to implement in an efficient way: as long as we maintain CT in Standard Cover Order, we can simply traverse over it and add those $X \in CT$ to $\mathcal{C}$ for which $X = (t \setminus C) \cap X$. This way, covering a single transaction t with CT has a worst case complexity of $O(|CT|)$.

4.2 Covering with Overlap

Here we introduce the main contribution of this section, dubbed SNOR. Although the standard cover function has been empirically shown to work well, it is limited to finding non-overlapping covers, which results in overly large code tables. Although intuitively appealing, overlap is a more subtle issue than it may seem at first. Or, as Peter Grünwald asked us a long time ago: "If you are after the minimal description, why describe an item more than once?"

Let us start with an example where overlap makes sense. Consider a transaction $t = \{ABCDE\}$ and a code table consisting of patterns ABC, BCD, and CDE with high usage and singletons with low usage. COVER happily describes t using ABC and singletons D and E. While correct, this cover is much more costly in terms of bits than when we simply use ABC and CDE.

Next, we illustrate why we should not expect too much from overlap. Consider a dataset where we plant ABC, BCD, and CDE with overlap. It seems obvious that these are the only patterns we need to most succinctly describe the data. This is correct, *iff* they are strictly independent. The moment the empirical frequency of a *combination* of these deviates from the marginals, it will quickly becomes beneficial to include that combination (e.g., $ABCD$) into the pattern set. That is, overlap *only* provides a gain in terms of MDL if patterns are (sufficiently) independent, but otherwise there is not much difference. This means it is unlikely that by allowing overlap, we will see much better compression or much smaller code tables. What it will do is prevent combinations of independent and overlapping patterns to be unnecessarily included in the code table. In short, overlap should lead to (somewhat) smaller code tables.

Covering with overlap is, however, not as simple as removing the condition $X \cap C = \emptyset$ from COVER. To illustrate what goes wrong, suppose we have a transaction $t = ABCDEF$ and a code table consisting of patterns ABC, BCD, CDE, and DEF, all with approximately equally high usage, and singletons with low usage. COVER without overlap returns $C = \{ABC, DEF\}$, which is a good result. It is concise and has a short encoded length $L(t)$. Obviously, we would like to obtain the same, or similarly good result if we do allow for overlap. If we simply remove the overlap condition from COVER, however, it returns all four patterns in the code table, which is highly redundant (C is described thrice!) and has a much longer encoded length. We encountered this problem when developing KRIMP [36], but found that straightforward ideas—such as a alternate orders— or more complicated ones—such as penalising doubly-described items—did not solve this problem. Focusing on fixed cover orders for efficiency, we did not realise that COVER resembles a greedy algorithm for weighted set cover *without overlap*, and that by replacing it with an efficient greedy algorithm for weighted set cover *with overlap*, we would have obtained the solution we were after.

That is, the key idea to making overlap work is to follow the underlying idea of COVER to iteratively select that $X \in CT$ that describes the most *uncovered* elements of t in (likely) the fewest number of bits. To this end, SNOR starts just like COVER with an empty cover, $C = \emptyset$. It then iteratively chooses and adds that $X \in CT$ such that $X \subseteq t$ and $|t \setminus C|$ is maximal. That is, SNOR iteratively adds that code table element that covers most of the items in the transaction that have not been covered yet. Further, it break ties by using the same Standard Cover Order as COVER, preferring X of higher cardinality, then those with higher support, and finally, determining preference lexicographically as to prefer those patterns with (the highest probability of having) a short encoded length.

SNOR is a bit harder to implement efficiently than COVER. To cover a given transaction t, we lazily maintain a list L of patterns $X \in CT$ that can still be added to cover C, i.e., for which $X \subseteq t$ and $X \not\subseteq C$. We order the elements $X \in L$ by how many uncovered items of t they can cover, i.e., $|(t \cap X) \setminus C|$. This way we can quickly access those $X \in CT$ that cover the most uncovered elements of t. Covering a single transaction t with CT then has a worst case complexity of $O(|C| \cdot |CT|)$. Luckily it is much faster in practice.

5   Search Algorithms

We will use the original KRIMP and SLIM algorithms for finding good tables and experiment with both cover functions, COVER and SNOR, to assess the impact of covering with or without overlap.

5.1   KRIMP

The KRIMP algorithm was introduced by Siebes et al. [30,36] for mining good code tables given a set of candidate patterns. We give the pseudo-code of KRIMP as Algorithm 1. KRIMP starts with the singleton code table ST (line 1) and a candidate collection $\mathcal{F}$ of frequent itemsets up to a given *minsup*. The candidates are ordered first descending on support, second descending on itemset length, and third lexicographically. Each candidate X is considered in turn by inserting it in CT (1.3) which we maintain ordered first descending on itemset length, second descending on support, and third lexicographically. We accept a candidate only if it improves compression (1.4). If accepted, we post-prune the code table by reconsidering all elements $X \in CT$ w.r.t. their contribution to compression (1.5). Post-pruning is a pivotal step as it helps to weed out patterns that are no longer useful.

Algorithm 1: The KRIMP Algorithm [36]

    input   : Database D and Candidate Set $\mathcal{F}$
    output : Code Table CT
1 $CT \leftarrow ST$;
2 **foreach** $X \in \mathcal{F}$ in **Standard Candidate Order do**
3      $CT' \leftarrow CT \cup X$;
4   **if** $L(D, CT') < L(D, CT)$ **then**
5         $CT \leftarrow post\text{-}prune(CT')$;
6 **return** CT;

5.2 SLIM

The SLIM algorithm was introduced by Smets and Vreeken [32] for mining good code tables directly from data. We give the pseudo-code of SLIM as Algorithm 2. SLIM starts with the singleton-only code table ST (line 1). In every iteration (line 2) we consider all pairwise combinations of $X, Y \in CT$ and estimate the gain in compression, $L(D, CT) - \hat{L}(D, CT \oplus Z)$, if we were to add $Z = X \cup Y$ to the current code table. Iteratively, in decreasing order of estimated gain, we add a candidate Z to CT (line 3), cover the data, and compute the actual gain (line 4). If compression improves, we accept the candidate, otherwise we reject it. If accepted, we post-prune by reconsidering weeding out every pattern in CT that no longer contributes towards compression (line 5). We then update the candidate list (line 2), and continue until no candidate decreases the total compressed size, after which we are done.

Algorithm 2: The SLIM Algorithm [32]

    input   : Database D
    output : Code Table CT
1 $CT \leftarrow ST$;
2 **foreach** $Z \in \{X \cup Y : X, Y \in CT\}$ in **Gain Order do**
3      $CT' \leftarrow CT \cup Z$;
4   **if** $L(D, CT') < L(D, CT)$ **then**
5         $CT \leftarrow post\text{-}prune(CT')$;
6 **return** CT;

Table 1. Datasets. Given are the numbers of transactions ($|D|$), items ($|I|$), resp. classes ($|C|$), the minimal support threshold ($minsup$) such that KRIMP finishes within a reasonable time, and the encoded size using the singleton-only code table ($L(D, ST)$).

Dataset	$|D|$	$|I|$	$|C|$	$minsup$	$L(D, ST)$
Adult	48842	97	2	20	3569724
Anneal	898	71	5	1	62827
Breast	699	16	2	1	27113
Chess (k-k)	3196	75	2	1200	687120
Chess (kr-k)	28056	56	18	1	1083046
DNA Amplification	4590	392	-	1	212640
Heart	303	50	5	1	20543
Hepatitis	155	52	-	1	14959
Iris	150	19	3	1	3058
Led7	3200	24	10	1	107091
Letter Recognition	20000	102	26	50	1980244
Mushroom	8124	119	2	1	1111287
Nursery	12960	32	5	1	569042
Pageblocks	5473	44	5	1	216552
Pen Digits	10992	86	10	50	1140795
Pima	768	38	2	1	26250
TicTacToe	958	29	2	1	45977
Wine	178	68	3	1	14101

6 Experiments

In this section we empirically investigate the benefit of using SNOR instead of COVER in KRIMP [36] and SLIM [32]. For conciseness, we will write KRIMP and SLIM for their original setup using COVER. We write KRAMP and SLAM to refer to their respective setup using SNOR, where the 'a' comes from 'overlap'.

6.1 Setup

We evaluate the algorithms on 18 datasets that together cover a wide range of characteristics. We give the base statistics in Table 1. We use the largest and most dense databases from the LUCS/KDD dataset repository, and obtain *Chess (kr-kp)* and *Chess (kr-k)* from the UCI repository. As real data we consider the DNA Amplification database [25], which contains DNA copy number amplifications. Such copies activate oncogenes and are hallmarks of advanced tumours.

KRIMP and KRAMP require a set of frequent patterns as candidates. For these, we mine closed frequent patterns with a minimum support threshold ($minsup$) set such that they finish within an hour. We give these minsup values in Table 1. SLIM and SLAM permit setting a $minsup$, but by default use $minsup$

Table 2. Kramp *mines better pattern sets than* Krimp. For 18 benchmark datasets, we show the number of transactions ($|D|$), number of items ($|I|$), and give the number of discovered patterns ($|CT|$, lower is better) and the total encoded size ($L(D, M)$, lower is better) for Krimp resp. Kramp using the *minsup* values given in Table 1. We also give the relative differences between the number of discovered patterns and total encoded sizes. Green is better, red is worse. On average, Kramp mines smaller code tables that compress better than those mined by Krimp.

Dataset	$|D|$	$|I|$	Krimp $|CT|$	$L(D, M)$	Kramp $|CT|$	$L(D, M)$	diff $\Delta|CT|$	$\Delta L(D)$
Adult	48842	97	1346	907803	1043	870096	−22.5%	−4.2%
Anneal	898	71	141	22873	131	21128	−7.1%	−7.6%
Breast	699	16	32	4896	37	4921	8.8%	0.5%
Chess (kr-kp)	3196	75	213	293176	149	280824	−30.0%	−4.2%
Chess (kr-k)	28056	56	1721	667339	1386	643654	−19.5%	−3.5%
DNA Amp.	4590	392	615	79829	586	78454	−4.7%	−1.7%
Heart	303	50	99	12241	107	11829	8.1%	−3.4%
Hepatitis	155	52	92	8317	89	8146	−3.3%	−2.1%
Iris	150	19	22	1475	23	1487	4.5%	0.8%
Led7	3200	24	168	30664	161	31310	−4.5%	2.1%
LetRecog	20000	102	1317	832483	1138	856730	−13.6%	2.9%
Mushroom	8124	119	268	274220	376	242017	40.3%	−11.7%
Nursery	12960	32	262	258898	302	270634	15.3%	4.5%
Pageblocks	5473	44	75	10967	71	10934	−5.3%	−0.3%
Pen Digits	10992	86	1072	523805	811	546219	−24.3%	4.3%
Pima	768	38	84	9199	84	9171	0.0%	−0.3%
TicTacToe	958	29	181	28880	172	28285	−5.0%	−2.1%
Wine	178	68	117	10965	123	10809	5.1%	−1.4%
Average			435	221002	377	218147	−3.2%	−1.5%

1, which is what we use unless stated differently. For all algorithms, we enable pruning the code table after every accepted new pattern.

We implemented all algorithms in C++ and make our code available for research purposes.[5] To compare fairly, we use equally (un)optimised implementations of the cover functions[6] and report wall-clock running times. All experiments were conducted as single-threaded runs on a Windows 11 machine with an AMD Ryzen 5 3600X processor (3.8GHz) and 32 GB of memory.

---

[5] https://vreeken.eu/prj/snor/.

[6] We use the more general 'cpoul' branch of cover functions, rather than the heavily optimised 'cccp' branch that does not allow overlap.

6.2   Compression

First, we investigate the effect of SNOR in terms of the models that KRIMP and
SLIM find. In particular, we are interested in whether allowing overlap leads to
better compression. To this end we run each method on all datasets. We show
the results for KRIMP and KRAMP in Table 2, and for SLIM and SLAM in Table 3.

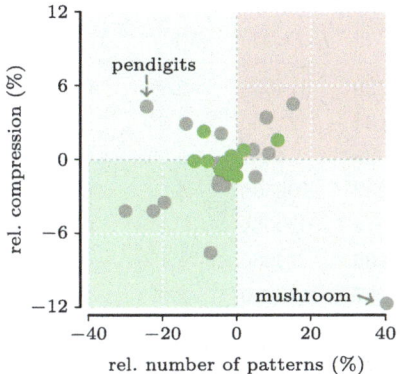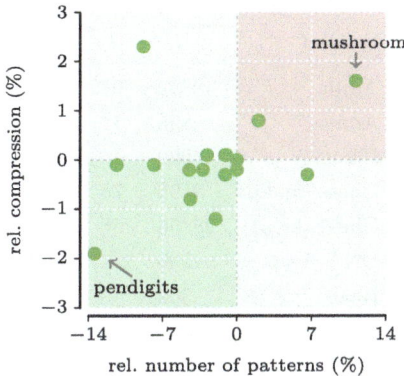

Fig. 1. SLAM *and* KRAMP *mine smaller code tables.* On the left, we consider KRIMP
versus KRAMP (grey) and SLIM versus SLAM (green) using *minsup* values such that
KRIMP finishes within reasonable time (see Table 1). On the right, we consider SLIM
and SLAM using *minsup* = 1. The bottom left quadrants contain datasets where overlap leads to *smaller* pattern sets that compress *better*. The bottom right those where
overlap leads to *larger* pattern sets that compress *better*, the top left *smaller* pattern
sets that compress *worse*, and the top right *larger* pattern sets that compress *worse*.
With only few exceptions, KRAMP and SLAM provide a benefit over KRIMP and SLIM.

Before we discuss these results in detail, we consider Fig. 1. There, we plot
the relative gain in compression, i.e.,

$$\Delta L(D) = \frac{L_{\text{COVER}}(D, CT) - L_{\text{SNOR}}(D, CT)}{L_{\text{COVER}}(D, CT)} ,$$

against the relative number of discovered patterns, i.e.,

$$\Delta|CT| = \frac{|CT_{\text{COVER}}| - |CT_{\text{SNOR}}|}{|CT_{\text{COVER}}|} ,$$

between using KRIMP or SLIM with COVER resp. SNOR. In both cases, larger
negative numbers are better.

On the left, we plot the difference between KRIMP and KRAMP, resp. SLIM
and SLAM, using the *minsup* thresholds from Table 1. We see that for the majority of datasets, SNOR leads to smaller pattern sets that compress better. The
effects of SNOR are more pronounced for KRIMP than for SLIM. For *Mushroom*,
KRAMP mines a larger code table (40% more patterns) that compresses much

Table 3. SLAM *mines smaller pattern sets than* SLIM. For 18 benchmark datasets, we show the number of transactions ($|D|$), number of items ($|I|$), and give the number of discovered patterns ($|CT|$, lower is better) and the total encoded size ($L(D, M)$, lower is better) for SLIM resp. SLAM for $minsup = 1$. We also give the relative differences between the number of discovered patterns and total encoded sizes. Green is better, red is worse. On average, SLIM with SNOR compresses as well as vanilla SLIM, but does so with smaller code tables.

Dataset	$|D|$	$|I|$	SLIM $|CT|$	SLIM $L(D, M)$	SLAM $|CT|$	SLAM $L(D, M)$	diff $\Delta|CT|$	diff $\Delta L(D)$
Adult	48842	97	1264	813778	1229	814599	−2.8%	0.1%
Anneal	898	71	147	20679	134	21146	−8.8%	2.3%
Breast	699	16	28	4251	28	4251	0.0%	0.0%
Chess (kr-kp)	3196	75	243	102730	232	102568	−4.5%	−0.2%
Chess (kr-k)	28056	56	1056	622946	1044	623665	−1.1%	0.1%
DNA Amp.	4590	392	613	78933	606	78660	−1.1%	−0.3%
Heart	303	50	99	10210	97	10089	−2.0%	−1.2%
Hepatitis	155	52	90	6798	86	6742	−4.4%	−0.8%
Iris	150	19	22	1489	22	1489	0.0%	0.0%
Led7	3200	24	147	29408	150	29648	2.0%	0.8%
LetRecog	20000	102	1635	664262	1527	666218	−6.6%	0.3%
Mushroom	8124	119	384	208608	427	211936	11.2%	1.6%
Nursery	12960	32	207	247390	205	247742	−1.0%	0.1%
Pageblocks	5473	44	77	10911	71	10898	−7.8%	−0.1%
Pen Digits	10992	86	1409	452528	1220	443946	−13.4%	−1.9%
Pima	768	38	76	8342	76	8327	0.0%	−0.2%
TicTacToe	958	29	133	23090	118	23074	−11.3%	−0.1%
Wine	178	68	124	10676	120	10652	−3.2%	−0.2%
Average			431	184279	411	184203	−3.1%	0.0%

better (12% better), while for *Pen Digits* it mines a smaller code table (24% fewer patterns) that compresses a bit worse (4.3%).

On the right, we plot the difference between SLIM and SLAM when we set the minimal support threshold to 1. We see again that for the majority of datasets, SNOR leads to smaller pattern sets that compress better. The effect is less pronounced than for KRIMP, with improvements up to 14% fewer patterns and up to 2% better compression. Interestingly, we find that SLAM performs much worse on *Mushroom* than SLIM; the former returns a much larger code table (13% more patterns) that compresses worse (1.5% more bits). For *Pen Digits*, however, we see an improvement, as SLAM finds a code table that is much smaller (14% fewer patterns) that compresses better (2% fewer bits). It is unclear why the relative compression gains for these two datasets are reversed between KRIMP and KRAMP resp. SLIM and SLAM.

Next, we consider the detailed results for KRIMP and KRAMP in Table 2. On average, KRAMP mines code tables that contain 3.2% fewer patterns yet compress 1.5% better. There is no clear trend, but it seems that datasets where *minsup* is set larger than 1, SNOR has a stronger effect.

Next, we consider the results of SLIM and SLAM in Table 3. We here find that overlap leads to smaller code tables (3.1% on average) that compress as well as those without overlap. We attribute this to the advanced search scheme that SLIM employs, which enables it to mine candidates from how patterns are being used to cover the data. It is unclear why exactly overlap has such a strongly negative effect for *Mushroom*. We postulate that this also has to do with the SLIM search scheme, in that it initially finds overlapping 'building blocks' and then refines these into variations of a theme, rather than discovering new themes.

6.3   Classification

Next, we evaluate the effect of SNOR on how well the mined code tables generalise. To this end, we consider the task of classification and follow the setup of Van Leeuwen et al. [21], in which we classify an unseen transaction t as label $l \in L$, according to $\arg\min_{l \in L} L(t \mid CT_l)$, where CT_l is the code table inferred for D_l. As data, we consider 15 benchmark datasets. As SLIM has been shown to outperform KRIMP in terms of classification [32], we focus on SLIM and SLAM.

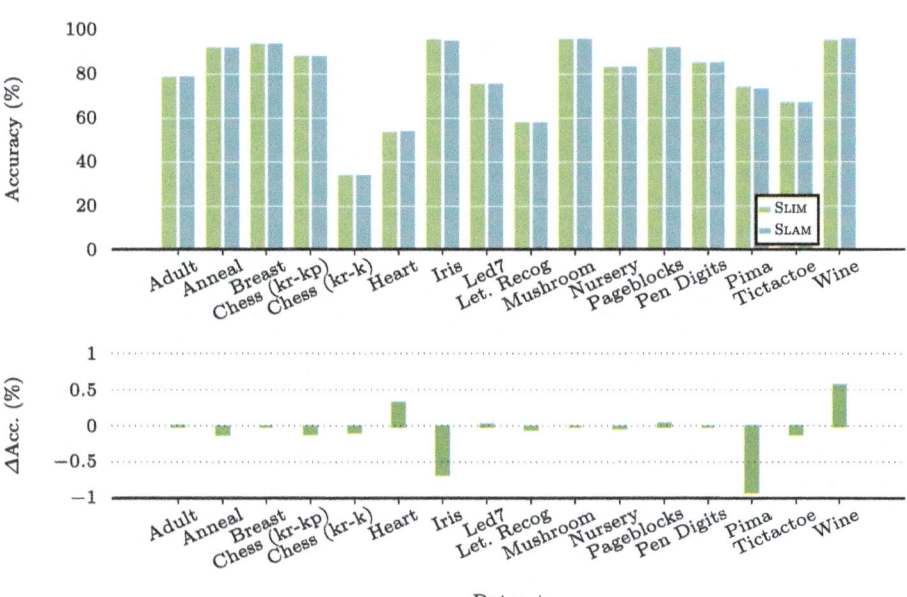

Fig. 2. SLAM *performs almost as well as* SLIM *in classification.* We plot classification accuracies (top) and absolute differences in accuracy (bottom) for SLIM and SLAM. For many datasets there is little to no difference in accuracy.

Table 4. SLAM *performs almost as well as* SLIM *in classification.* For sixteen benchmark datasets, we show the number of transactions ($|D|$), number of items ($|I|$), number of classes ($|C|$), and average accuracy (acc) with standard deviation (std) as obtained using 10-fold cross validation using SLIM resp. SLAM. Bold numbers are better.

Dataset	$|D|$	$|I|$	$|C|$	SLIM acc ±	std	SLAM acc ±	std	diff
Adult	48842	97	2	78.40 ±	0.63	**78.41** ±	0.52	0.01%
Anneal	898	71	5	**91.54** ±	2.17	91.43 ±	3.37	−0.11%
Breast	699	16	2	**93.28** ±	3.83	**93.28** ±	2.63	0.00%
Chess (kr-kp)	3196	75	2	**87.83** ±	2.03	87.73 ±	2.86	−0.10%
Chess (kr-k)	28056	56	18	**33.70** ±	0.61	33.62 ±	0.75	−0.08%
Heart	303	50	5	53.14 ±	9.34	**53.47** ±	11.18	0.33%
Iris	150	19	3	**95.33** ±	4.27	94.67 ±	8.33	−0.66%
Led7	3200	24	10	75.22 ±	2.87	**75.25** ±	2.36	0.03%
Letter Recog.	20000	102	26	**57.59** ±	1.20	57.55 ±	1.04	−0.04%
Mushroom	8124	119	2	**95.51** ±	0.67	**95.51** ±	0.57	0.00%
Nursery	12960	32	5	**82.91** ±	0.91	82.89 ±	0.87	−0.02%
Pageblocks	5473	44	5	91.65 ±	1.02	**91.69** ±	0.82	0.04%
Pen Digits	10992	86	10	**84.93** ±	1.03	**84.93** ±	0.95	0.00%
Pima	768	38	2	**73.70** ±	6.09	72.79 ±	2.88	−0.91%
TicTacToe	958	29	2	**66.81** ±	5.94	66.70 ±	4.90	−0.11%
Wine	178	68	3	94.94 ±	5.42	**95.51** ±	3.36	0.57%
Avg. Acc				**78.53** ±	3.00	78.46 ±	2.96	−0.07%
Avg. Rank				**1.41**		1.59		

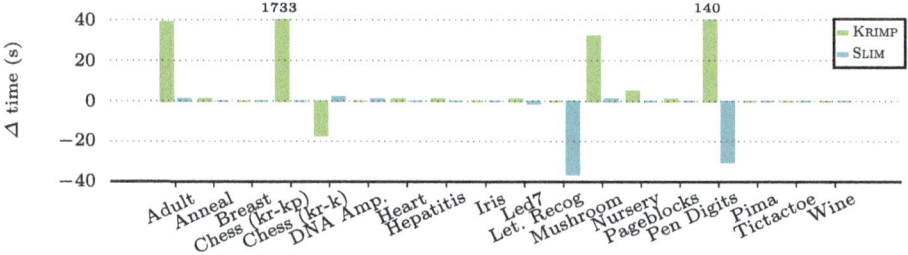

Fig. 3. SNOR *slows down* KRIMP *but speeds up* SLIM. We plot the difference in runtime between KRIMP and KRAMP, resp. SLIM and SLAM. For many datasets there is little to no difference in runtime. For SLIM, we find cases with significant speed-up, while for KRIMP we find the opposite.

We report the average accuracy over 10-fold cross-validation in Fig. 2 and Table 4. We see that both methods perform very similarly. If we count carefully, we find that SLIM beats SLAM in 8 cases, SLAM beats SLIM in 5, but then also find that the differences in accuracy are very small (0.07% on average) and well within the standard deviations. The results on the very small *Iris* and

Pima datasets skew the average somewhat; if we disregard these, SLAM in fact outperforms SLIM with a difference of 0.03% (77.68 ± 2.69 versus 77.71 ± 2.58).

6.4   Runtime

Next, we consider the effect of SNOR in terms of runtime. In Fig. 3, we show the differences in runtime of KRIMP and KRAMP, resp. SLIM and SLAM. We see that for most datasets, the overhead incurred by SNOR is negligible. For KRAMP we have four notable exceptions: it is much slower on *Adult*, *Chess (kr-kp)*, *Mushroom*, and *Pen Digits*. For SLAM we have two notable exceptions: it is much faster than SLIM on *Letter Recognition* and *Pen Digits*. It is unclear whether this is because these datasets focus on a computer vision task.

7   Conclusion

With the goal of obtaining smaller pattern sets that describe the data even better, we studied how to enable KRIMP and SLIM to describe the data using sets of patterns that overlap. We studied the original COVER function used by both, and saw that it is an efficient but crude approximation of greedy weighted set cover. We saw that because of the interplay between usage and code lengths an optimal solution is infeasible, and that naively extending COVER to allow overlap leads to bad results. Taking inspiration from the aspects that make COVER work well, we then proposed SNOR, an efficient heuristic for succinctly describing a transaction using overlapping patterns. The key idea is that rather then considering the patterns in a fixed order, we iteratively select that pattern that covers the most uncovered items.

Through an extensive set of experiments, we showed that KRIMP and SLIM obtain better results with SNOR than without. On average, SNOR leads to smaller pattern sets that compress the data at least as well, permit equally good classification, and in case of SLIM even speeds up the inference of the code tables. In other words, pattern mining is simply better with SNOR.

While the experiments show that SNOR leads to smaller pattern sets, whether these patterns are also simpler is left for future investigation. We postulate that SNOR speeds SLIM up exactly because of this reason, i.e., once SLIM has found the (overlapping) building blocks *ABC*, *CDE*, *EFG* that make up the data, thanks to SNOR it can accurately estimate if these are used (sufficiently) independently to expect any gain in compression by combining them; this, as opposed to SLIM having to consider combining *ABC* with *D*, and then *ABCD* with *E*, etc. to describe the data equally well without overlap.

We focused on KRIMP and SLIM, by which it remains a question whether SNOR would also improve DIFFNORM, and related, it is interesting to investigate variants of SNOR for sequential or graph data. Another challenge for the future is to how to enable KRIMP and SLIM to effectively deal with destructive noise.

All in all, even after 20 years plenty of opportunities to revisit remain.

Acknowledgments. The authors thank Arno Siebes for suggesting to use MDL for pattern mining, the many great discussions that followed, but above all on delivering on the 'fun guarantee' he gave them at the start of their Ph.D. trajectories.

References

1. Aggarwal, C.C., Han, J. (eds.) Frequent Pattern Mining. Springer (2004)
2. Akoglu, L., Tong, H., Vreeken, J., Faloutsos, C.: COMPREX: compression based anomaly detection. In: Proceedings of the 21st ACM Conference on Information and Knowledge Management (CIKM), Maui, HI. ACM (2012)
3. Bariatti, F., Cellier, P., Ferré, S.: GraphMDL: graph pattern selection based on minimum description length. In: Berthold, M.R., Feelders, A., Krempl, G. (eds.) IDA 2020. LNCS, vol. 12080, pp. 54–66. Springer, Cham (2020). https://doi.org/10.1007/978-3-030-44584-3_5
4. Bertens, R., Vreeken, J., Siebes, A.: Keeping it short and simple: summarising complex event sequences with multivariate patterns. In: Proceedings of the 22nd ACM International Conference on Knowledge Discovery and Data Mining (SIGKDD), San Francisco, CA, pp. 735–744 (2016)
5. Bertens, R., Vreeken, J., Siebes, A.: Efficiently discovering unexpected pattern co-occurrences. In: Proceedings of the SIAM International Conference on Data Mining (SDM 2017). SIAM (2017)
6. Budhathoki, K., Vreeken, J.: The difference and the norm — characterising similarities and differences between databases. In: Appice, A., Rodrigues, P.P., Santos Costa, V., Gama, J., Jorge, A., Soares, C. (eds.) ECML PKDD 2015. LNCS (LNAI), vol. 9285, pp. 206–223. Springer, Cham (2015). https://doi.org/10.1007/978-3-319-23525-7_13
7. Cover, T.M., Thomas, J.A.: Elements of Information Theory. Wiley-Interscience, New York (2006)
8. Cueppers, J., Vreeken, J.: Below the surface: summarizing event sequences with generalized sequential patterns. In: Proceedings of the ACM International Conference on Knowledge Discovery and Data Mining (SIGKDD) (2023)
9. Farag, I.: Efficient data summarization with overlapping patterns. Msc. thesis, Saarland University (2018)
10. Fischer, J., Oláh, A., Vreeken, J.: What's in the box? Exploring the inner life of neural networks with robust rules. In: Proceedings of the International Conference on Machine Learning (ICML). PMLR (2021)
11. Galbrun, E.: The minimum description length principle for pattern mining: a survey. Data Min. Knowl. Disc. **36**(5), 1679–1727 (2022)
12. Grünwald, P.: The Minimum Description Length Principle. MIT Press (2007)
13. Grünwald, P., Roos, T.: Minimum description length revisited. Int. J. Math. Ind. **11**(1) (2019)
14. Hedderich, M., Fischer, J., Klakow, D., Vreeken, J.: Label-descriptive patterns and their application to characterizing classification errors. In: Proceedings of the International Conference on Machine Learning (ICML). PMLR (2022)
15. HHeikinheimo, H., Siebes, A., Vreeken, J., Mannila, H.: Low-entropy set selection. In: Proceedings of the 9th SIAM International Conference on Data Mining (SDM), Sparks, NV, pp. 569–580 (2009)
16. Kersten, R., Siebes, A.: A structure function for transaction data. In: Proceedings of the SIAM International Conference on Data Mining (SDM) (2011)

17. Kersten, R., Siebes, A.: Smoothing categorical data. In: Proceedings of the European Conference on Machine Learning and Principles and Practice of Knowledge Discovery in Databases (ECML PKDD) (2012)
18. Koopman, A., Siebes, A.: Characteristic relational patterns. In: Proceedings of the 15th ACM International Conference on Knowledge Discovery and Data Mining (SIGKDD), Paris, France, pp. 437–446 (2009)
19. Koutra, D., Kang, U., Vreeken, J., Faloutsos, C.: VoG: summarizing graphs using rich vocabularies. In: Proceedings of the SIAM International Conference on Data Mining (SDM), Philadelphia, PA, pp. 91–99. SIAM (2014)
20. van Leeuwen, M., Siebes, A.: STREAMKRIMP: detecting change in data streams. In: Proceedings of the European Conference on Machine Learning and Principles and Practice of Knowledge Discovery in Databases (ECML PKDD), Antwerp, Belgium, pp. 672–687 (2008)
21. van Leeuwen, M., Vreeken, J., Siebes, A.: Compression picks the item sets that matter. In: Proceedings of the 10th European Conference on Principles and Practice of Knowledge Discovery in Databases (PKDD), Berlin, Germany, pp. 585–592 (2006)
22. van Leeuwen, M., Vreeken, J., Siebes, A.: Identifying the components. Data Min. Knowl. Disc. **19**(2), 173–292 (2009)
23. Li, M., Vitányi, P.: An Introduction to Kolmogorov Complexity and its Applications. Springer (1993)
24. Makhalova, T., Kuznetsov, S.O., Napoli, A.: Mint: MDL-based approach for Mining INTeresting numerical pattern sets. Data Min. Knowl. Disc. **36**(1), 108–145 (2022)
25. Myllykangas, S., Himberg, J., Böhling, T., Nagy, B., Hollmén, J., Knuutila, S.: DNA copy number amplification profiling of human neoplasms. Oncogene **25**(55), 7324–7332 (2006)
26. Proenca, H.M., van Leeuwen, M.: Interpretable multiclass classification by mdl-based rule lists. Inf. Sci. **512**, 1372–1393 (2020)
27. Rissanen, J.: Modeling by shortest data description. Automatica **14**(1), 465–471 (1978)
28. Sampson, O., Berthold, M.R.: Widened krimp: better performance through diverse parallelism. In: Proceedings of Advances in Intelligent Data Analysis (IDA) (2014)
29. Shah, N., Koutra, D., Zou, T., Gallagher, B., Faloutsos, C.: Timecrunch: interpretable dynamic graph summarization. In: Proceedings of the ACM International Conference on Knowledge Discovery and Data Mining (SIGKDD), pp. 1055–1064. ACM (2015)
30. Siebes, A., Vreeken, J., van Leeuwen, M.: Item sets that compress. In: Proceedings of the 6th SIAM International Conference on Data Mining (SDM), Bethesda, MD, pp. 393–404. SIAM (2006)
31. Smets, K., Vreeken, J.: The odd one out: identifying and characterising anomalies. In: Proceedings of the 11th SIAM International Conference on Data Mining (SDM), Mesa, AZ, pp. 804–815. Society for Industrial and Applied Mathematics (SIAM) (2011)
32. Smets, K., Vreeken, J.: SLIM: directly mining descriptive patterns. In: Proceedings of the 12th SIAM International Conference on Data Mining (SDM), Anaheim, CA, pp. 236–247. Society for Industrial and Applied Mathematics (SIAM) (2012)
33. Tatti, N., Vreeken, J.: The long and the short of it: Summarizing event sequences with serial episodes. In: Proceedings of the 18th ACM International Conference on Knowledge Discovery and Data Mining (SIGKDD), Beijing, China, pp. 462–470. ACM (2012)

34. van Leeuwen, M., Galbrun, E.: Association discovery in two-view data. IEEE Trans. Knowl. Data Eng. **27**(12), 3190–3202 (2015)

35. Vreeken, J., van Leeuwen, M., Siebes, A.: Characterising the difference. In: Proceedings of the 13th ACM International Conference on Knowledge Discovery and Data Mining (SIGKDD), San Jose, CA, pp. 765–774 (2007)

36. Vreeken, J., van Leeuwen, M., Siebes, A.: KRIMP: mining itemsets that compress. Data Min. Knowl. Disc. **23**(1), 169–214 (2011)

37. Wallace, C.S.: Statistical and Inductive Inference by Minimum Message Length. Springer (2005)

38. Wiegand, B., Klakow, B., Vreeken, J.: Mining easily understandable models from complex event data. In: Proceedings of the SIAM Conference on Data Mining (SDM). SIAM (2021)

39. Yang, L., van Leeuwen, M.: Truly unordered probabilistic rule sets for multi-class classification. In: Amini, M.R., Canu, S., Fischer, A., Guns, T., Kralj Novak, P., Tsoumakas, G. (eds.) ECML PKDD 2022. LNCS, vol. 13717. Springer, Cham (2022). https://doi.org/10.1007/978-3-031-26419-1_6

World Cup of Flags: A Multi-relational Competitive Vexillology Dataset from Twitter

Wouter Duivesteijn[1](✉), Ad Feelders[2], and Arno Knobbe[3]

[1] Technische Universiteit Eindhoven, Eindhoven, The Netherlands
w.duivesteijn@tue.nl
[2] Universiteit Utrecht, Utrecht, The Netherlands
a.j.feelders@uu.nl
[3] Universiteit Leiden, Leiden, The Netherlands
a.j.knobbe@liacs.leidenuniv.nl

Abstract. We introduce a new dataset, chronicling the World Cup of Flags, a competitive vexillology tournament held on Twitter. The dataset combines challenges arising from three angles. Firstly, the data is multi-relational, so analysis techniques need to be able to respect that; for instance, conclusions on prior probabilities must be drawn across one-to-many or many-to-many relations spanning several tables. Secondly, the data stems from a tournament composed of a group phase followed by a knockout phase; assessing performance of a specific competitor needs to incorporate the relative strength of the opponents gleaned from incomplete data: most flags will not meet most other flags in the tournament. Finally, this competition was held on Twitter; as a consequence it spiraled completely out of control. An auxiliary contribution of this paper is the downright bizarre story of precisely how the World Cup of Flags unfolded, including ideological differences between vexillological, maximalist, and nationalist voting blocs, a takeover by a substantial wave of Zimbabwean Twitter personalities, and involvement of both the Prime Minister and Leader of the Opposition of Trinidad & Tobago. Hence, the World Cup of Flags dataset is a publicly available benchmark of noisy data, concerning matches in a tournament structure that is familiar from many sports, also encompassing multi-relational data mining challenges.

Keywords: Public Dataset · Multi-relational Data · Tournament Analysis · Vexillology · Twitter

1 Introduction

In times of a pandemic, you have to make your own fun. During the COVID-19 pandemic, public life ground to a standstill. Many sports leagues big and small, national and international, stopped playing as well. This deprived an already relatively bored world population of a much-needed dose of entertainment. While some scratched their itch by seeking out those rare competitions that weren't

M. van Leeuwen and J. Vreeken (Eds.): Arno Siebes Festschrift, LNCS 16067, pp. 75–93, 2026.
https://doi.org/10.1007/978-3-032-03028-3_5

suspended (such as the Belarusian football league [29]), the team behind Election Maps UK channeled their lack-of-sports-related boredom in a more original manner. This organization normally focuses on producing maps illustrating electoral outcomes, but as a side project, they started a new Twitter account called World Cup of Flags [37].

The concept was at best very loosely defined, probably deliberately: every match in the World Cup of Flags would be a Twitter poll with two options, each option being the emoji of a flag of a country. There were no further instructions. People would vote for their favorite of the two flags, the best-designed of the two flags, or just the flag of their own country, and none of this would be against any rules because there essentially were no rules. The result was a tournament that became a glorious sprawling mess.

For the casual observer, there were clear patterns in the tournament. For instance, the more colorful, exuberant flag designs did quite well, and the more standard tribands (three equally sized horizontal or vertical stripes and nothing else) did quite badly. Hence, if one were to analyze the data from the World Cup of Flags, it is quite possible that one could draw strong correlations between flag design elements and tournament performance. However, this task is not as straightforward as it seems, for the following reasons:

- the World Cup of Flags was a tournament with 192 entrants, starting with a six-team group phase; the top-two of each group would qualify for a straight knockout phase. This makes it difficult to define an appropriate measure of success: most flags never meet each other during the tournament, not all flags play an equal number of matches, design elements success measurement must correct for the success of design elements of competing flags, etcetera;
- flag characteristics, flag performance, tournament stage characteristics, and match statistics are best modeled in separate tables of a multi-relational database, and some of these tables have many-to-many connections. Hence, one should approach this problem with techniques from Multi-relational Data Mining [5,9,15]; the obvious step to propositionalize [16,17,19] the four tables into a single table should be done with care, as one might mess with prior probabilities;
- the contest is very ill-defined, and has no historical precedent. Hence, no historical data on flags or design elements is available, which makes it impossible to use the win prediction techniques that are standard in sports analytics [18,23,31,32].

The main contribution of this paper is the *World Cup of Flags dataset*, manually collected from publicly available information on Twitter, detailing the results of the World Cup of Flags (WCoF) tournament. Publicly available multi-relational datasets are relatively scarce (although a UCI ML repository equivalent for relational data does exists [26], albeit on a much smaller scale), and even scarcer are publicly available datasets that combine challenges from multi-relational data mining and sports analytics. Hence, the WCoF dataset provides a benchmark for developing better methods on the crossroads of multi-relational data mining and analyzing a tournament encompassing a group phase and knockout phase, in the absence of any historical performance data. The dataset can be downloaded by anyone at [6] and from the companion website at http://wouterd.win.tue.nl/WCoF/.

2   World Cup of Flags Tournament Synopsis

No matter how well an event is planned, it is almost inevitable that some things will go wrong. The World Cup of Flags, however, was loosely planned from the start, and what exactly was being assessed in individual matches was left vague. As a result, ad hoc repairs to Twitter-exacerbated excesses needed to be deployed. This section chronicles the main events. Part of this section is reconstructed from memory and aggregation from observed Twitter replies, so references are scarce.

2.1   Original Setup

A collection of 192 countries were selected to participate in the tournament, to be distributed into 32 groups of 6 flags each. This already poses a small problem: the United Nations has 193 member states, on top of which the WCoF organization decided to include the two non-member observer states Holy See and Palestine. To reduce these 195 countries down to 192 participating flags, the WCoF organization decided to let six countries participate on three joint tickets [3]. Two of those made sense: Monaco and Indonesia have the exact same flag, as do Chad and Romania. The third joint ticket was Luxembourg and the Netherlands, who both fly a horizontal triband of red, white, and blue, albeit different shades of blue. This proved to be slightly controversial with Dutch Twitter, since the flags of Luxembourg and the Netherlands are about as similar as the flags of Italy and the Republic of Ireland, but those flags were participating on separate, individual tickets. In the end it hardly mattered: the joint ticket finished dead last in Group P, which is hardly surprising given that it is a boring triband, and failing jointly is probably preferable over failing individually.

Each group was held on a single day, and organized as a single round robin. This meant that fifteen matches were held per group. After the matches were held, final group placement was decided by number of wins; in case of ties, vote percentage was employed as tie breaker; this was only required to distance the group winner from the runner-up in Group M, while a total order by number of wins was achieved in all other groups. In the Group Q match between Azerbaijan and Portugal, the organization accidentally displayed the Azerbaijani flag in both options of the Twitter poll. Portugal won the match anyway, and neither flag survived the group stage, so the effect of this mistake on the tournament outcome is likely negligible.

The top-two flags from every group progressed to the knockout stage, which was a straight knockout. In the first knockout round, group winners were drawn against runners-up from other groups; subsequent rounds were drawn randomly. Results of the matches, final group standings, and subsequent knockout phase bracket were aggregated on a website [35]. However, this website misses the results from Zimbabwe matches that were later replayed with other teams, due to the disqualification which we will discuss in Sect. 2.5.

2.2   Emergence of Several Types of Voters

Within the first few matches, discussions broke out in reply threads to the individual match tweets between voters. Clear groups of voters emerged that shared some ideas about what is desirable in a flag, but those ideas were markedly different between the groups. Three types of votes can be clearly delineated.

Firstly, we have the *vexillological voters*. These are people who have a deep attachment to the concept 'flag', in all its uses. To these voters, a good flag must be distinctive from other flags, and recognizable, even from a distance. After all, flags are often observed from a distance, and you still would want to be able to recognize what country the flag represents. As a consequence, these voters prefer distinctive yet simple designs, with not too many elements, not too many distinct colors, clear contrasts between elements, etcetera. This type of voter is likely to appreciate a flag like the one of Switzerland: a white cross in the center of a red field.

Secondly, we have the *maximalist voters*. These are people who like to have a palette of cheerful colors (the more the merrier), elements that people typically do not associate with flags, and wildlife. After all, there isn't a flag in the world that cannot be improved by adding an extra kiwi bird firing a laser beam from its eye [28]. This type of voter is likely to appreciate a flag like the one of Kiribati: a red sky over white and blue waves with a yellow frigate bird flying over a rising sun with seventeen rays. Maximalist voters are perennially at odds with vexillological voters, since their principles steer them in opposite directions: maximalists prefer more on their flags, but vexillologists find that less is more.

Thirdly and finally, we have the *nationalist voters*. These are people who think that the flag of their country is the best, since it is the flag of their country and therefore it logically follows that flag is the best flag there is. This logic is hard to disprove, but it is also very likely to be met with disagreement from both other voter types.

It is perhaps important to reiterate at this point that the World Cup of Flags came with no instructions or clarifications. There were Twitter polls with two flag emoji as options, but no question was asked. It is therefore impossible to say that any of the three voter groups necessarily had the wrong idea.

2.3   Repelling the Takeover by Zimbabwean Twitter

During the group phase, in most groups, roughly the same number of voters would vote on all group matches. This somewhat makes sense: when you're interested in the contest, you participate in all matches that run concurrently. Strange things happened in Group C, however, where Brazil and Kenya took the top-two spots for the majority of the day. In the final hour, however, Zimbabwe suddenly rallied to first place. When the dust had settled, total vote counts for all ten Group C matches *not* involving Zimbabwe ranged from 1 954 to 2 145, while total vote counts for all five Group C matches involving Zimbabwe ranged from 2 267 to 2 389. If we would leave out a single outlier in both categories, all other matches involving Zimbabwe received at least 200 more votes than all other matches not involving Zimbabwe. This ten percent boost was *just* enough to push Zimbabwe up to first place and eliminate Kenya from the tournament. The WCoF organizer tweeted as explanation that a Zimbabwean football team had become involved in the World Cup of Flags, and recruited all their friends to vote for the Zimbabwean flag. This was considered to be part of the fun and games of the contest, and so the group phase rolled on.

The situation polarized in the knockout phase. Several Zimbabwean Twitter personalities rallied their followers to vote for the Zimbabwean flag, which boosted the number of votes for the Zimbabwean contest in the Rounds of 64 and 32. This started to annoy vexillological, maximalist, and casual voters: among those groups, negative sentiments were voiced about the large group of voters who were not interested in the World Cup of Flags as a whole, but only voted on the matches involving Zimbabwe. The WCoF organizer chose to react, by announcing a new rule to be implemented at the start of the Round of 16: if a match would receive at least 1.5 times the total number of votes of another match on the same day, then the winning flag would be disqualified [36]. Even though nationalistic voters from Zimbabwe did not violate any rule so far, apparently the organizer found that the spirit of the World Cup of Flags is best served by those voters who involve themselves with all matches (no matter which countries participate in them), which is behavior that this rule was meant to reward. It immediately caused more trouble.

2.4   Political Peace in Trinidad & Tobago

In the Round of 16, the flag of Zimbabwe was pitted against the flag of Trinidad & Tobago, where the World Cup of Flags had become quite popular too. The number of votes for matches involving Trinidad & Tobago did not differ much from the average number of votes for other matches in their group, but this changed in the knockout phase. In the Rounds of 64 and 32, vote totals of matches involving Trinidad & Tobago were eclipsed by vote totals for matches involving Zimbabwe, but elevated w.r.t. all other flags. As a result, when Zimbabwe and Trinidad & Tobago met, their respective Twitter spheres exploded.

While vote totals in Zimbabwe were driven by a football team and Twitter personalities, in Trinidad & Tobago serious politicians got involved. On the day of the Round of 16 match, Dr. Keith Rowley (the Prime Minister) and Kamla Persad-Bissessar (the Leader of the Opposition) both tweeted out a call to vote for Trinidad & Tobago in this match [10]. While Zimbabwe won preceding knockout rounds quite comfortably, this exceptional show of political unity made Trinidad & Tobago a serious challenger: the fight was very close throughout the day. Since a close match boosts individual efforts to rally more home support, the WCoF organizer chose to immediately change the rule that was only just announced: disqualification would only follow if the winning flag in the exceedingly popular match would also win over 60% of the votes in the match itself. This extra clause ensured that the winner of this Round of 16 would advance to the quarter finals; Zimbabwe progressed once more.

2.5   Zimbabwe's Disqualification and Subsequent Repechage

In the Quarter Final, the inevitable happened. The flag of Zimbabwe trounced the flag of Namibia, taking over 80% of the votes, but a significant number of people couldn't be bothered to vote on the other quarter final held on the same day. Hence, the flag of Zimbabwe was disqualified from the contest according to the updated rules. This led to a torrent of online abuse for the WCoF organizer.

A disqualification in a knockout tournament leaves a hole in the schedule, so a small repechage tournament was held: the flag that lost to Zimbabwe in the Round of 64 was pitted against the flag that lost to Zimbabwe in the Round of 32, the winner of that match played the flag that lost to Zimbabwe in the Round of 16, the winner of that match played the flag that lost to Zimbabwe in the quarter final, and the winner of that match took Zimbabwe's spot in the semi finals. The World Cup of Flags dataset includes both the original knockout matches involving Zimbabwe and the repechage matches held after disqualification; you can imagine that this complicates the drawing of a knockout tournament tree quite a bit.

3   The World Cup of Flags Dataset

The World Cup of Flags (WCoF) dataset is publicly available in xls format via the link on the companion website at http://wouterd.win.tue.nl/WCoF/, and at [6]. The dataset consists of four tables, each of which can be accessed in their own tab of the xls file. In the next section, we discuss the tables and their relations; in Sect. 3.2, we discuss some design choices made in filling out some characteristics of the tables and underlying rationales; in Sect. 3.3, we discuss the main learning task and how usage of information from multiple tables may make this more interesting.

3.1   Tables and Relations

The core entity in the WCoF dataset is the **Flag**, which is identified by its Alpha-3 code as defined in the ISO 3166 international standard. The **Flag** and its characteristics is the main topic of the first table: **FlagCharacteristics**.

FlagCharacteristics. The primary key of this table is the three-letter **Flag** code. The remaining columns of this table are simply characteristics of this flag, as outlined in Table 1.

Matches. Flags participate in matches of the World Cup of Flags. Each match is fought between two competing flags. Flags participate in an integer number of matches between 5 and 12, depending on their performance in the tournament as a whole: better performance leads to further advancement in the tournament, which results in more matches.

Table 1. Columns of the `FlagCharacteristics` table. The table lists all non-binary-valued columns, and specifies the form their values take. The table omits all binary-valued columns, whose values are either `Yes` or `No` (`HasHorizontal`, `HasVertical`, ...).

Column name	Values
`Flag`	ISO 3166 country code, Alpha-3 form (primary key)
`CountryName`	Free text
`ContinentRegion`	{Africa, ArabianPeninsula, Asia, Caribbean, Caucasus, CentralAmerica, Europe, NorthAmerica, Oceania, SouthAmerica}
`Shape`	{Rectangle, Square, TwoTriangles}
`MainDesign`	{Bands, Cross, DefacedBlueEnsign, Diagonal, DiagonalCross, Field, Other, SerratedFields, Triangles}
`MainDirection`	{Diagonal, Horizontal, None, Triangle, Vertical}
`NrBands`	$\{0, 1, \ldots, 14\}$
`NrMainDesignFeatures`	$\{0, 1, \ldots, 48\}$
`NrDistinctColors`	$\{0, 1, \ldots, 9\}$
`NrEmblemElements`	$\{0, 1, \ldots, 262\}$
`NrStars`	$\{0, 1, \ldots, 50\}$
`NrPeople`	$\{0, 1, 2\}$

A match between two flags can occur twice during the tournament: two flags that have met in the group phase but both advance to the knockout phase, can potentially meet again somewhere in that knockout phase. Therefore, the primary key for the `Matches` table consists of a combination of three columns: `Stage`, `Flag1`, and `Flag2`. Details on these and the remaining columns of this table can be found in Table 2.

Stages. A match is part of a stage, whose details are listed in a separate `Stages` table. A stage consists of all the World Cup of Flags matches that are held on the same day and in the same phase of the tournament. In the beginning of the tournament, a stage consists of all matches in a single group. The bigger rounds of the knockout phase took place over multiple days, which is why the Round of 64 (which is one single phase of the tournament) consists of several stages. Conversely, the full set of repechage matches (cf. Sect. 2.5) were held on the same day, but these matches belong to separate phases of the tournament. Details on the columns and their contents can be found in Table 3.

FlagPerformance. Based on the manner in which flags perform in individual matches, and the results of other matches in the same group, a flag may advance in the tournament or not. Since this information is nontrivial to aggregate from the other three tables, we gather it in a separate fourth table, named `FlagPerformance`. The three-letter flag code is the primary key of this table; details of all its columns can be found in Table 4.

Table 2. Columns of the `Matches` table: name and content value specification.

Column name	Values
Stage	{Group A, Group B, Group C, ..., Group Z, Group AA, Group AB, Group AC, Group AD, Group AE, Group AF, R64A, R64B, R64C, R64D, R32A, R32B, R32C, R32D, R16A, R16B, QFA, QFB, RepR32, RepR16, RepQF, SF, 3PP, F} (primary key)
Flag1	ISO 3166 country code, Alpha-3 form (primary key)
Flag2	ISO 3166 country code, Alpha-3 form (primary key)
Votes1	$[0, 1]$ (vote share of `Flag 1`)
Votes2	$[0, 1]$ (vote share of `Flag 2`)
Winner	ISO 3166 country code, Alpha-3 form
NrVotes	$\{0, 1, ..., 39\,431\}$
NrReplies	$\{0, 1, ..., 1\,000\}$
NrRetweets	$\{0, 1, ..., 3\,300\}$
NrLikes	$\{0, 1, ..., 2\,800\}$

Table 3. Columns of the `Stages` table: name and content value specification.

Column name	Values
Stage	{Group A, Group B, Group C, ..., Group Z, Group AA, Group AB, Group AC, Group AD, Group AE, Group AF, R64A, R64B, R64C, R64D, R32A, R32B, R32C, R32D, R16A, R16B, QFA, QFB, RepR32, RepR16, RepQF, SF, 3PP, F} (primary key)
Date	YYYY-MM-DD
Day	{Mon, Tue, Wed, Thu, Fri, Sat, Sun}
Weekday	{Yes, No}
Phase	{Group, R64, R32, R16, QF, SF, 3PP, F}

3.2   Filling Tables: Design Choices and Rationales

Given a flag, it may be possible to fill out the `FlagCharacteristics` table in multiple ways. For example, the main design of the flag of Spain consists of three horizontal stripes, of which the middle (yellow) stripe is substantially thicker than the border (red) stripes. One could also view this as a field of yellow, with two red stripes superimposed. Purely by looking at the flag, one could choose *Bands* as `MainDesign`, *Horizontal* as `MainDirection`, and 3 as `NrBands`. Alternatively, one could choose *Field* as `MainDesign`, *None* as `MainDirection`, and 2 as `NrBands`.

Sometimes, choices made in such matters are purely in the eye of the beholder. Wherever possible, however, we adhered to local laws describing the flag. The Constitution of Spain [4] specifies that the Spanish flag consists of three stripes, so of the two alternatives mentioned in the previous paragraph, the first is correct. Conversely, Mauritania, whose flag follows a similar design structure,

Table 4. Columns of the `FlagPerformance` table: name and content value specification.

Column name	Values
`Flag`	ISO 3166 country code, Alpha-3 form (primary key)
`Group`	{A, B, C, ..., Z, AA, AB, AC, AD, AE, AF}
`SurvivedGroupPhase`	{Yes, No}
`GroupPlacement`	{1, 2, 3, 4, 5, 6} (lower number = higher placement = better)
`ReachedR32`	{Yes, No}
`ReachedR16`	{Yes, No}
`ReachedQF`	{Yes, No}
`ReachedSF`	{Yes, No}
`ReachedFinal`	{Yes, No}
`Won3rdPlacePlayoff`	{Yes, No}
`WonFinal`	{Yes, No}

has a law [1] stipulating that its flag is a field flanked by two stripes. Hence, of the two alternative descriptions, the second is correct for Mauritania. Following national laws, Japan's disc represents the Sun, while Palau's disc represents the Moon.

The `Matches` table contains tweet metadata such as number of likes and retweets; this data was collected two weeks after the corresponding tweet was released. If the numbers of replies, retweets, and likes had grown too large, they may have only been available in the form "1.1K", in which case we added padding zeros to the number. This solution is far from elegant, but such numbers are extreme outliers in the distribution anyway, so we doubt the need for further precision. The `Stages` table contains a column describing whether the corresponding matches took place on a weekday or weekend day; here, the weekend is taken to encompass Saturday and Sunday, even though we are aware that this is not applicable worldwide (which is likely irrelevant for the accuracy of drawn conclusions). The `FlagPerformance` table is straightforwardly filled, although counts are off due to the repechage (cf. Sect. 2.5): nine flags reached the quarter finals, and seventeen flags reached the round of sixteen. The `SurvivedGroupPhase` column can be trivially reconstructed from the `GroupPlacement` column: the former equals *Yes* if and only if the latter equals 1 or 2.

3.3 Learning Tasks

To the casual spectator, it is not that difficult to synthesize observations on which types of flags do well in the competition, and what binds those flags. For instance, the ubiquitous relatively boring triband designs typically do worse than more eclectic designs, while flags featuring an overly detailed and frilly emblem typically do not do so well. Therefore, the main question of the WCoF dataset is: *Can we find (combinations of) flag characteristics that correlate with performing exceptionally well or badly in the tournament?*

At first sight, answering this question seems straightforward. Flag characteristics are detailed in the FlagCharacteristics table (cf. Table 1), the performance of flags within the tournament is aggregated in the FlagPerformance tables (cf. Table 4), and there is a one-to-one relation between these tables with the same single column as primary key. Hence, the naive approach would be to join these two tables, and let one's favorite flat-table data mining technique go to town on this single flat table. There are, however, several complications that make this approach suboptimal.

Firstly, the goal for a flag is to achieve "good performance", but it is not apriori clear how to translate that concept into something data mining methods can understand. The World Cup of Flags is a tournament with a group phase, followed by a knockout phase consisting of several rounds. It is simple to indicate *factors* that correlate with better performance. We could for instance claim that a flag performed better than a competitor when it achieved a higher placement in the group, reached a later stage in the knockout phase, racked up more overall wins. One could select one of these as a target column for a classification approach, but already we are talking about three subtly distinct learning tasks here. How to extract a concept of "better performance" out of the information available in Table 4 is a question with more than one correct answer.

Secondly, the competitor matters. If two eclectic flag designs face off in a match against each other, only one can win. Suppose that the one eclectic flag beats another eclectic flag, and elsewhere a cross beats a triband. If we were to compute simple statistics over the number of wins, the reached knockout stage, or the group placement, we would draw the conclusion that the cross performs better than the eclectic flag, which performs better than the triband. Such conclusions would be premature, since we haven't incorporated the strength of the opposition in our workings. Therefore, if one tries to assess which flag characteristics lead to better performance, one needs to take into account the characteristics of the flag on the other side of the matchup. This information can be extracted by making our analysis also encompass data from the Matches table (cf. Table 2). However, relations with this table are many-to-many. Hence, in order to incorporate its data *without invalidating conclusions on prior probabilities*, one ought to treat the learning task as one from Multi-Relational Data Mining [8].

Finally, context matters. As we have discussed in Sect. 2.2, the World Cup of Flags attracted several types of voters. The distribution over these types is different under different conditions. Twitter users can see the progression of the vote as soon as they cast theirs. If a particular vote is tight, vexillologically-inspired voters will merely be more interested in the result, but nationalistically-inspired voters are more likely to act and convince their friends to vote. Hence, a tight vote tends to increase turnout and skew the distribution of voter types. These effects may be stronger or weaker depending on which part of the population has idle time on their hands (which will be different on distinct days of the week). Therefore, it is quite possible that the conclusions of our analysis become more precise if we incorporate such contextual information. Such information is contained in the Twitter-metadata columns of the Matches table (cf. Table 2), and in the Stages table (cf. Table 3).

4   Analyzing Matches

In this section, we analyze the data from a matches perspective. This means that we treat each row in the Matches table as an example. Since this table in its normalized form contains only limited information (a reference to the stage, two references to the flags involved, and some statistics about the outcome), we opt to denormalize the relations of Matches with its direct neighbors Stages and Flags. In other words, we join in details about the matches and the two flags involved, which, since all relations are many to one (foreign key lookups), provides no obstacles.

In its simplest form, which we refer to as **denormalized 1**, the flag details of the two flags will appear as two separate sets of attributes in the denormalized Matches table, identified by postfixes '_1' and '_2'. Since in the actual competition, the order of the first and second flag also (potentially) matters, it makes sense to keep this order-specific representation. Although straightforward, there is a downside to this representation. This has to do with the expectation that human voters will likely compare the properties of flag 1 to those of flag 2, and having the current denormalized representation doesn't encourage such direct comparison. Direct comparison would be greatly helped by adding derived attributes that emphasise the differences between flags. This leads us to the second denormalized dataset, **denormalized 2**, which just has 'difference' attributes (aside from the target). When in the first denormalized dataset, a flag characteristic is represented in numeric form, e.g. `NrDistinctColors_1` and `NrDistinctColors_2`, **denormalized 2** simply has `diff_NrDistinctColors`. In the case of binary attributes (e.g. `HasTriangle_*`), these are first translated to a 0/1-representation.

4.1   Denormalized Data

Before comparing the two representations, let's first analyze **denormalized 1** using Subgroup Discovery. Our SD workhorse for this paper is the Leiden-developed tool SubDisc [24,25], available as a GUI-based Java implementation, as well as a Python and R package[1]. We'll look for subgroups for which the target (**winner**) has value 'Flag 1', which holds for 52.56 % of the cases (in other words, there is a slight bias towards the first flag mentioned in the match). We'll look for subgroups of limited depth that show an interesting and significant deviation from the prior distribution. For the latter, we use the concept of the *Distribution of False Discoveries* obtained through swap-randomized experiments [7]. To encourage significance, we set a subgroup's minimum size to 10 matches (although most quality measures encourage much larger subgroups anyway).

We start with a depth-1 and depth-2 analysis using the default quality measure *SubDisc Quality*, defined as

$$SQ(s, t) = \frac{\text{WRAcc}(s, t)}{p_t(1 - p_t)}$$

---

[1] https://github.com/SubDisc.

Table 5. Six simple subgroups.

Id	Coverage	Quality	Posterior	Conditions
1	329	0.213	0.614	MainDesign_2 = 'Bands'
2	356	0.197	0.601	NrMainDesignFeatures_2 ≤ 4
3	260	0.186	0.623	NrBands_1 ≤ 2
4	355	0.179	0.594	NrBands_2 ≥ 2
5	424	0.177	0.582	HasTriangle_2 = 'No'
6	134	0.173	0.701	HasTriangle_1 = 'Yes'

Table 6. Pattern team of size 3 (out of 13). The joint entropy is $H(3, 8, 13) = 2.995$.

Id	Coverage	Quality	Posterior	Conditions
3	260	0.186	0.623	NrBands_1 ≤ 2
8	277	0.165	0.606	MainDirection_2 = 'Horizontal'
13	284	0.145	0.595	HasBlue_1 = 'Yes'

where

$$\mathrm{WRAcc}(s, t) = p_{st} - p_s \cdot p_t.$$

and p_s is the proportion of the subgroup, p_t is the proportion of positive cases, and p_{st} is the proportion of positive cases within the subgroup (all measured relative to the total dataset). This measure is thus order-equivalent with the well-known WRAcc [20], but uses scaling to obtain values between $[-1, 1]$, compared to the arbitrary and dataset-dependent subinterval of $[-0.25, 0.25]$ for WRAcc. At depth 1, the DFD procedure determines significant subgroups ($\alpha = 0.05$) to require a SubDisc Quality of at least 0.137.

SubDisc finds 13 subgroups, of which the best six are presented in Table 5. This suggests that flag 1 wins if flag 2 has a bands-design and/or few design features. Furthermore, these results suggest that having few bands and/or having a triangle are a positive characteristic. Due to the clumsy order-based representation, similar properties of either flag are identified, so we can already expect **denormalized 2** to be a more natural representation. The best subgroup at this depth covers about 60 % of the cases and has a posterior probability of 0.614, an 8.8 percentage point increase. Flag 1 having a triangle provides a bigger absolute increase (subgroup 6), but the size of this subgroup decreases to 24.5 %. As a group, the subgroups produce a convex hull in ROC space whose Area Under the Curve is 0.652, not impressive, but reasonable given the shallow search depth.

Since there appears to be some redundancy in the found subgroups, we compute a pattern team [13] based on maximizing the joint entropy of the subgroups [14, 22]. The three subgroups listed in Table 6 are selected for diversity among the 13 significant subgroups. The joint entropy approaching the maximum of 3 bits of information indicates that these three findings are nearly independent (while still being predictive).

Table 7. Rules with Combined Conditions

Id	Coverage	Quality	Posterior	Conditions
1	277	0.282	0.664	MainDesign_2 = 'Bands' $\wedge$ HasTriangle_2 = 'No'
2	262	0.274	0.668	MainDesign_2 = 'Bands' $\wedge$ NrMainDesignFeatures_2 $\leq$ 5
3	278	0.271	0.658	NrBands_2 $\geq$ 2 $\wedge$ NrMainDesignFeatures_2 $\leq$ 5
4	298	0.260	0.644	NrBands_2 $\geq$ 2 $\wedge$ HasTriangle_2 = 'No'

Table 8. Subgroups based on flags-difference attributes, depth 1.

Id	Coverage	Quality	Posterior	Conditions
1	320	0.278	0.643	NrMainDesignFeatures $\geq$ 0
2	450	0.194	0.584	HasTriangle $\geq$ 0
3	195	0.173	0.646	NrBands $\leq$ -1
4	480	0.159	0.571	HasDiagonal $\geq$ 0
5	401	0.156	0.579	HasBlue $\geq$ 0
6	412	0.150	0.575	HasHorizontal $\leq$ 0
7	346	0.133	0.578	NrDistinctColors $\geq$ 0

Table 9. Subgroups based on flags-difference attributes, depth 2.

Id	Coverage	Quality	Posterior	Conditions
1	231	0.327	0.719	NrMainDesignFeatures $\geq$ 0 $\wedge$ HasHorizontal $\leq$ 0
2	289	0.324	0.678	NrMainDesignFeatures $\geq$ 0 $\wedge$ HasDiagonal $\geq$ 0
3	259	0.315	0.691	NrMainDesignFeatures $\geq$ 0 $\wedge$ NrBands $\leq$ 2
4	310	0.309	0.661	NrMainDesignFeatures $\geq$ 0 $\wedge$ HasArabicScript $\leq$ 0
5	289	0.302	0.668	NrMainDesignFeatures $\geq$ 0 $\wedge$ HasTriangle $\geq$ 0

A depth-2 run results in 305 subgroups, of which the first 4 are shown in Table 7. Due to the overlap in some features, and the possibility to combine conditions in different ways, already quite a bit of redundancy is showing up, which continues down the result list. Subgroups at depth 3 don't provide much improvement over these depth-2 results. The AUC of the ROC convex hull of these results is 0.691.

Switching to **denormalized 2**, with just difference attributes, we find 7 significant subgroups at depth 1 (the minimum quality threshold for significance is recomputed whenever the data or the search parameters change). They are listed in Table 8. Right off the bat, we note that the best difference-based attribute is more predictive than any of the flag-specific attributes (SubDisc Quality 0.278 vs. 0.213), despite some flag-specific attributes scoring reasonably well. Going one level deeper, we get the (truncated) list of results in Table 9.

For comparing the two data representations, we present descriptive statistics about each run in Table 10. Clearly, the second representation is more powerful.

Table 10. Representation comparison.

representation	AUC ($d = 1$)	AUC ($d = 2$)	best ($d = 1$)	best ($d = 2$)
denormalized 1	0.652	0.691	0.213	0.282
denormalized 2	0.658	0.714	0.278	0.327

Table 11. Summary of the selected logistic regression model.

Coefficient	Estimate	Pr($> |\mathbf{z}|$)	$\exp(\hat{\beta})$
(Intercept)	0.07940	0.435	
HasHorizontal	-0.70935	< 0.001	0.49
HasDiagonal	0.73586	0.004	2.09
HasTriangle	0.57658	0.006	1.78
NrMainDesignFeatures	0.06031	< 0.001	1.06
HasRed	-1.09875	< 0.001	0.33
HasGreen	-0.85405	< 0.001	0.43
HasBrown	-1.97085	< 0.001	0.14
NrDistinctColors	0.61189	< 0.001	1.84
HasWrittenText	-0.90331	0.003	0.41
HasHat	-1.84687	0.002	0.16

A combination of the two datasets might produce even better results. This is left as an exercise to the user.

4.2 A Global Model

In this section, we build a global model to predict the outcome of matches. For ease of interpretation we restrict ourselves to logistic regression, and only use numerical differences as features (**denormalized 2**). To select a model, we first fit a logistic regression model using all features except for NrPeople, because it was very strongly correlated to HasPerson ($r = 0.95$). This model was used as the initial model in a stepwise search using the Bayesian Information Criterion for scoring models. In each step a feature could be added to or removed from the model.

The selected model is summarized in Table 11. We can read that having horizontal bands is unfavorable, and about halves the winning odds, whereas having a diagonal division is favorable and approximately doubles the odds. These results are in line with the SD results described in Table 8. The colors red, green, and (especially) brown are to be avoided, but in general more distinct colors increases the odds by 84% for each additional color. We can also use the model to try and answer the question whether it is an advantage to be displayed on the left side of the screen. For two identical flags (at least with respect to the model features), all feature values (numerical differences) are zero, and the probability that flag 1 wins is given by:

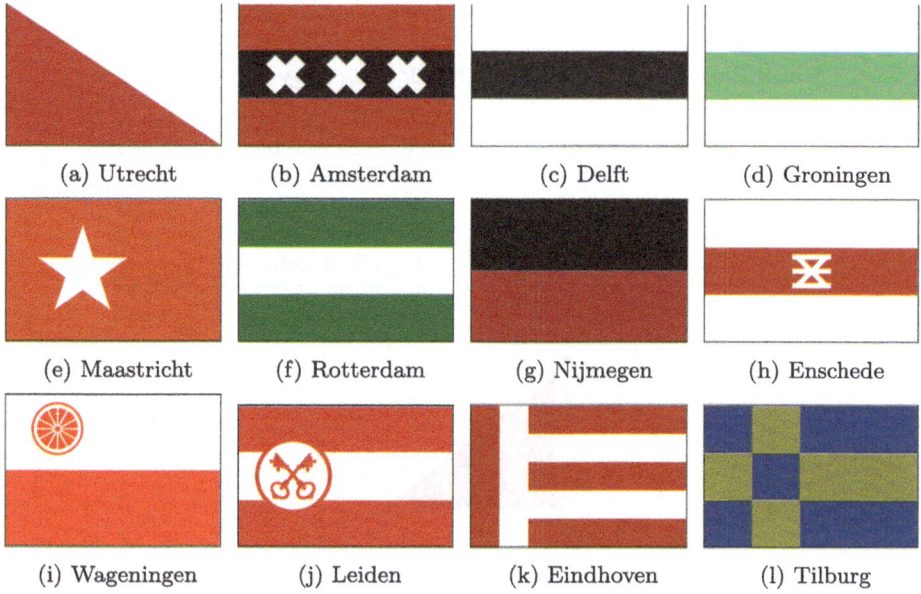

Fig. 1. Twelve flags of the university competition.

$$\Pr(Y = 1) = \frac{e^{\hat{\beta}_0}}{1 + e^{\hat{\beta}_0}} = \frac{e^{0.0794}}{1 + e^{0.0794}} = 0.52.$$

This seems to point to a slight advantage to be displayed on the left, but as we can see in Table 11 the intercept estimate is not significantly different from zero ($p = 0.435$), so there is not sufficient evidence for such an advantage.

To illustrate a possible application of the global model, we use it to run a single round-robin tournament between the flags of the 12 Dutch university cities. The flags are displayed in Fig. 1. For each match, the logistic regression model gives us a probability that the first flag will win. The outcome of the match is determined by tossing an appropriately biased coin. The final standings of the tournament are reported in Table 12.

Utrecht only lost to Tilburg in a match in which Utrecht was the clear favorite (win probability of 67%), but Tilburg won against the odds. The reader will note that Rotterdam against Groningen must have been a close call. Rotterdam had the slight advantage of being displayed on the left, which it managed to convert into a win. Figure 2 shows the winning flag in action. We will leave it as an exercise for the reader to determine the reasons for Utrecht's success.

5   Related Work

5.1   Multi-relational Data Mining and Datasets

For several subsets of data mining and machine learning tasks, well-known dataset repositories exist. For standard flat-table classification and regression

Table 12. Final standings of the tournament between Dutch universities.

Rank	Flag	City	Points	Rank	Flag	City	Points
1		Utrecht	10	7		Tilburg	6
2		Amsterdam	7	8		Enschede	5
3		Delft	7	9		Leiden	4
4		Groningen	7	10		Wageningen	4
5		Nijmegen	6	11		Maastricht	3
6		Rotterdam	6	12		Eindhoven	1

Fig. 2. The winning flag of the city of Utrecht in action.

tasks, the UCI machine learning repository [12] maintains 585 datasets. For frequent itemset mining, 12 datasets are available from the FIMI repository [2,11]. The MULAN repository [34] encompasses 26 datasets for multi-label classification and 18 for multi-target regression. For graph datasets, SNAP [21] provides 120 large network datasets, and the generically-named Network Data Repository [33] contains thousands of networks. The closest equivalent for multi-relational data mining is the Relational Dataset Repository [26], encompassing 73 datasets. Seven of these (BasketballMen, BasketballWomen, Chess, Hockey, NBA, NCAA, PremiereLeague (sic)) concern sports; they typically do not combine a group phase and knockout phase in a single tournament, and often information is available on player characteristics or prior tournament performance.

5.2   Sports Analytics

The learning tasks of predicting the outcomes and analyzing the factors influencing success in a competition are well-studied in the field of sports analytics. A body of work is dedicated to predicting the outcomes of the competition. Robberechts and Davis [31] compare result-based Elo ratings and goal-based Offense Defense Model ratings to predict the outcomes of the FIFA World Cup using ordered logit regression and bivariate Poisson model: the ordered logit model with Elo ratings as a independent variable outperforms the other models.

Many research papers in the field of sports analytics describe different methods of calculating the likelihood of winning at any given point in a game—

in-game win probability. Such models are based on features that capture the estimated pre-game rating/strength difference between the teams, as well as the features describing the current in-game situation. Robberechts et al. [32] discuss Bayesian statistical model predicting the running win, tie, and loss probabilities for the home team using data from four seasons of the major European football competitions. Lock and Nettleton [23] use a different model—Random Forest—to estimate win probability before each play of an NFL game. Pelechrinis [30] applies logistic regression model with 10 predictors—field position, down, ball possession time, etc.—to forecast the home team's running win probability.

Finally, one can be interested in looking for similarities between the teams—in the given context, flags—and ranking them. Muller and Langer [27] present Soccer TEam VEctors for learning real-valued vectors for soccer teams that places similar teams close to each other in the resulting vector space. These vectors can later be used as independent variables in various predicting models. Another rating system—Elo ranking—is based on team's past victories or defeats and the estimated strength of the opponent. It is an "earned rating system" meaning that the ranking is updated after each match [18].

6   Conclusions

We introduce the World Cup of Flags (WCoF) dataset, a publicly available dataset of information crawled from a competitive vexillology tournament held on Twitter. The dataset combines challenges from two specific subfields of data mining: on the one hand, the data is multi-relational in nature, on the other hand, properly analyzing the results of a tournament consisting of a group phase followed by a knockout phase is a fundamental sports analytics research topic. Add to that the inherent volatility of letting a tournament unfold through Twitter, and an intentionally loosely-defined objective of individual matches which allowed for polemic discourse, and we end up with a situation where the dataset shows seventeen flags qualifying for the Round of 16, the Prime Minister and the Leader of the Opposition of Trinidad & Tobago getting involved in an unusual show of political unity, and a horde of angry Zimbabwean Twitter users demanding the head of the tournament organizer on a plate.

We provide some initial example analyses on the WCoF dataset, using traditional propositional data mining techniques. From that, we learn that the colors brown and red, having written text, and having horizontal bands are the main flag characteristics disliked by people. Having triangles or a diagonal design, on the other hand, leads to a better performance. However, these observations should be taken as naive initial conclusions; proper multi-relational data mining is likely to find more and more interesting results.

Eventually, the World Cup of Flags was won by the only country flag on Earth without any of the colors red, white, and blue; the Dutch authors of this paper find this a discouraging finding. On the positive side, the (simulated) competition between the flags of Dutch university cities was a glorious victory for Utrecht, in honor of our professor emeritus.

Acknowledgments. We thank Ekaterina Meeuwesen, Nguyen Diep Linh, and Nils Hugo van der Grinten for their contributions to an early draft of Sect. 4 of this paper.

References

1. Agence Mauritanienne d'Information: L'Assemblée nationale adopte le projet de loi portant description du drapeau de la RIM (2017). http://fr.ami.mr/Depeche-42367.html. Accessed 26 Mar 2025
2. Bayardo Jr., R.J., Goethals, B., Zaki, M.J.: FIMI '04, Proceedings of the IEEE ICDM Workshop on Frequent Itemset Mining Implementations, Brighton, UK, 1 November 2004. CEUR Workshop Proceedings 126, CEUR-WS.org (2005)
3. Beaton: Of the World Cup of Flags. https://becomingthemuse.net/2020/05/19/of-the-world-cup-of-flags/. Accessed 26 Mar 2025
4. Constitución Española, Preliminary Part, Article 4.1. In: Agencia Estatal Boletín Oficial del Estado (1978)
5. De Raedt, L.: Logical and Relational Learning. Springer (2008)
6. Duivesteijn, W., Feelders, A.J., Knobbe, A.: World Cup of Flags dataset. figshare. Dataset (2025). https://doi.org/10.6084/m9.figshare.28869563
7. Duivesteijn, W., Knobbe, A.: Exploiting false discoveries — statistical validation of patterns and quality measures in subgroup discovery. In: 11th International Conference on Data Mining, pp. 151–160 (2011)
8. Džeroski, S.: Multi-relational data mining: an introduction. SIGKDD Explor. **5**(1), 1–16 (2003)
9. Džeroski, S., Lavrač, N. (eds.): Relational Data Mining. Springer (2001)
10. Fraser, N.: Politicians unite for Twitter World Cup of National Flags. In: Trinidad and Tobago Newsday, edition of Wednesday 13 May 2020. https://newsday.co.tt/2020/05/13/politicians-unite-for-twitter-world-cup-of-national-flags/. Accessed 26 Mar 2025
11. Goethals, B., Zaki, M.J.: FIMI '03, frequent itemset mining implementations. In: Proceedings of the ICDM 2003 Workshop on Frequent Itemset Mining Implementations, 19 December 2003, Melbourne, Florida, USA. CEUR Workshop Proceedings 90, CEUR-WS.org (2003)
12. Kelly, M., Longjohn, R., Nottingham, K.: The UCI Machine Learning Repository. https://archive.ics.uci.edu. Accessed 26 Mar 2025
13. Knobbe, A.J., Ho, E.K.Y.: Pattern teams. In: Fürnkranz, J., Scheffer, T., Spiliopoulou, M. (eds.) PKDD 2006. LNCS (LNAI), vol. 4213, pp. 577–584. Springer, Heidelberg (2006). https://doi.org/10.1007/11871637_58
14. Knobbe, A., Ho, E.K.Y.: Maximally informative k-itemsets and their efficient discovery. In: Proceedings of the 12th ACM SIGKDD International Conference on Knowledge Discovery and Data Mining (KDD), pp. 237–244 (2006)
15. Knobbe, A.: Multi-relational data mining. IOS Press (2006)
16. Kramer, S., Pfahringer, B., Helma, C.: Stochastic propositionalization of non-determinate background knowledge. In: Proceedings of ILP, pp. 80–94 (1998)
17. Krogel, M.-A., Rawles, S.A., Zelezn'y, F., Flach, P.A., Lavrač, N., Wrobel, S.: Comparative evaluation of approaches to propositionalization. In: Proceedings of ILP, pp. 197–214 (2003)
18. Lasek, J., Szlávik, Z., Bhulai, S.: The predictive power of ranking systems in association football. Int. J. Appl. Pattern Recognit. **1**(1) (2013). https://doi.org/10.1504/IJAPR.2013.052339

19. Lavrač, N., Džeroski, S., Grobelnik, M.: Learning nonrecursive definitions of relations with LINUS. In: Proceedings of EWSL, pp. 265–281 (1991)
20. Lavrač, N., Flach, P., Zupan, B.: Rule evaluation measures: a unifying view. In: Džeroski, S., Flach, P. (eds.) ILP 1999. LNCS (LNAI), vol. 1634, pp. 174–185. Springer, Heidelberg (1999). https://doi.org/10.1007/3-540-48751-4_17
21. Leskovec, J., Krevl, A.: SNAP Datasets: Stanford Large Network Dataset Collection (2014). http://snap.stanford.edu/data. Accessed 26 Mar 2025
22. van Leeuwen, M., Knobbe, A.: Diverse subgroup set discovery. Data Min. Knowl. Disc. **25**, 208–242 (2012)
23. Lock, D., Nettleton, D.: Using random forests to estimate win probability before each play of an NFL game. In: Machine Learning and Data Mining for Sports Analytics Conference Papers (2019)
24. Meeng, M., Knobbe, A.: Flexible enrichment with cortana — software demo. In: Proceedings of BeneLearn, pp. 117–119 (2011)
25. Meeng, M., Knobbe, A.: For real: a thorough look at numeric attributes in subgroup discovery. Data Min. Knowl. Disc. **35**(1), 158–212 (2021)
26. Motl, J., Schulte, O.: The CTU Prague Relational Learning Repository. arXiv preprint arXiv:1511.03086 (2015)
27. Muller, R., Langer, S.: Soccer team vectors. In: Machine Learning and Data Mining for Sports Analytics Conference Papers (2019)
28. New Zealand Ministry for Culture and Heritage: Laser Kiwi flag (2021). https://nzhistory.govt.nz/media/photo/fire-lazar. Accessed 26 Mar 2025
29. Parshakov, P.: The Spread of COVID-19 and Attending Football Matches: Lesson from Belarus. Unreviewed preprint. https://ssrn.com/abstract=3764404. Accessed 26 Mar 2025
30. Pelechrinis, K.: iWinRNFL: A Simple, Interpretable and Well-Calibrated In-Game Win Probability Model for NFL (2018). arXiv preprint https://arxiv.org/abs/1704.00197
31. Robberechts, P., Davis, J.: Forecasting the FIFA world cup – combining result- and goal-based team ability parameters. In: Brefeld, U., Davis, J., Van Haaren, J., Zimmermann, A. (eds.) MLSA 2018. LNCS (LNAI), vol. 11330, pp. 16–30. Springer, Cham (2019). https://doi.org/10.1007/978-3-030-17274-9_2
32. Robberechts, P., Van Haaren, J., Davis, J.: Who will win it? An in-game win probability model for football. In: Machine Learning and Data Mining for Sports Analytics Conference Papers (2019)
33. Rossi, R.A., Ahmed, N.K.: The network data repository with interactive graph analytics and visualization. In: Proceedings of AAAI, pp. 4292–4293 (2015)
34. Tsoumakas, G., Spyromitros-Xioufis, E., Vilcek, J., Vlahavas, I.: Mulan: a java library for multi-label learning. J. Mach. Learn. Res. **12**, 2411–2414 (2011)
35. World Cup of Flags results and fixtures. https://challonge.com/WorldCupOfFlags. Accessed 26 Mar 2025
36. World Cup of Flags disqualification rule tweet. https://twitter.com/WorldCupOfFlags/status/1259793793915781121. Accessed 12 May 2021
37. World Cup of Flags Twitter account. https://twitter.com/worldcupofflags. Accessed 12 May 2021

Learning and Reasoning

A Learning and Reasoning Manifesto

Luc De Raedt[1,4(✉)] [iD], Fredrik Heintz[2] [iD], Kristian Kersting[3] [iD],
and Giuseppe Marra[1] [iD]

[1] KU Leuven, Leuven, Belgium
{luc.deraedt,giuseppe.marra}@kuleuven.be
[2] Linköping University, Linköping, Sweden
fredrik.heintz@liu.se
[3] TU Darmstadt, hessian.AI, and DFKI, Darmstadt, Germany
kersting@cs.tu-darmstadt.de
[4] Örebro University, Örebro, Sweden

> *Only when Artificial Intelligence can rival*
> *Arno Intelligence will it truly succeed.*
> *But as Arno manifestly learns and reasons,*
> *so too must Artificial Intelligence.*
> *This manifesto is dedicated to that pursuit,*
> *and to Arno — who is inspiringly*
> *intelligent.*

Abstract. This manifesto provides a perspective on artificial intelligence in which the ability to both learn and reason are central. It then examines the state-the-of-the-art in artificial intelligence while focusing on foundation models and neurosymbolic artificial intelligence.

1 Introduction

This is the era of machine learning. Breakthroughs in machine learning have led to impressive tools such as large language models (LLMs) that are known, if not used, by almost everybody today. Unfortunately, the surprising results obtained through generative AI, rooted in machine learning, have also led to the false beliefs that AI is (almost) solved, that generative AI or more broadly (deep) machine learning is all what is needed, and that just feeding AI with more data and using larger models will lead to human-level intelligence.

And yet, despite these spectacular results, examples keep on appearing as to where purely data-driven approaches fail. Consider, for instance, the 'Alice in Wonderland' puzzle [1]:

"Alice has N brothers and she also has M sisters. How many sisters does Alice's brother have?"

When presenting this puzzle to many large language models (for several values of N and M), they return the incorrect answer M. However, if you add the

M. van Leeuwen and J. Vreeken (Eds.): Arno Siebes Festschrift, LNCS 16067, pp. 97–108, 2026.
https://doi.org/10.1007/978-3-032-03028-3_6

fact that Alice is a girl, then at least some of the models get the answer correctly. This is an example of a simple common sense fact that would be easy to add or infer in a knowledge-based approach.

Another example is that of the boat and the goat [2]:

> "A man and a goat are on one side of the river. They have a boat. How can they both go across?"

The answer of ChatGPT was:

> "The man takes the goat across the river first, leaving the boat on the original side. Then, the man returns alone with the boat. He leaves the goat on the other side and takes the boat back to the original side. Finally, he takes the cabbage across the river."

This is mixing up the nature of the puzzle it has to solve with a famous brain-teaser involving a boat, a goat, a wolf and a cabbage, and constraints on which it has received training data [2].

Both examples clearly illustrate the limitations of purely data-driven AI approaches and of large language models. They are good at pattern matching, but understand neither the contents of the patterns nor the context in which they are applied. They therefore do not properly reason [10–12]. This is clear in the boat-and-goat example, where a pattern learned on a famous complex puzzle is retrieved and applied to a much simplified variant of the original problem, yielding nonsensical results. This indicates that LLMs perform a kind of memorisation and re-use answers to previously seen problems to new situations, with variable success.

Purely data-driven approaches often fall short because they cannot fully capture the subtle edge cases, exceptions, and complex patterns that characterise real-world domains. This limitation highlights why relying solely on data may lead to incorrect or incomplete results. After all, data reflects specific instances, while rules represent general knowledge that typically holds across many or all cases. Actually, even current data-driven methods—whether supervised or self-supervised—are already grounded in human knowledge. In supervised learning, correct labels are provided by humans who apply their understanding of domain rules (e.g., traffic laws). In self-supervised learning, although no explicit labeling is used, models still depend on patterns found in data that was originally generated by humans, thus indirectly encoding human normative knowledge. Existing models, therefore, attempt to rediscover rules from data, which is of limited value when those rules are already known. Not only does this risk reproducing rules in an imperfect or incorrect form, but it also fails to advance knowledge.

What we truly need are learning algorithms that go beyond what is already known and that can generate novel insights while respecting established domain knowledge. From the standpoint of Trustworthy AI, the hallmark of AI made in Europe, purely data-driven approaches are insufficient. Instead, we need systems that combine data with explicit knowledge and reasoning to provide stronger guarantees and deeper understanding.

Let us illustrate the idea with an example of a self-driving car that needs to decide whether it can go first. It should be able to correctly answer the types of questions asked during driving license tests. A typical question confronts the examinee with an image depicting a traffic situation and a multiple choice question. To answer this type of question, one needs to combine perception to determine where the relevant objects are in the scene (where are the cars, the pedestrians, the traffic signs, ...) and interpret the resulting situation using the rules of traffic.

A purely data-driven approach for answering such questions would need vast amounts of training data, while a purely knowledge-driven approach would require encoding a complete and consistent knowledge base. Both would be impractical and likely would not really work. The data-driven approach would require too much data, would have to relearn the rules of traffic, and cannot guarantee that the decision making process would respect the rules of traffic. A purely knowledge-driven approach cannot solve the problem either. It would not be able to solve the perception task, and it would also require encoding the knowledge in a model that would need to be complete and consistent, and would need to cover all possible situations. Knowing when there is enough knowledge and finding the best encoding are highly non-trivial tasks. Furthermore, some rules of traffic are soft and therefore difficult to encode in an unambiguous manner (e.g., the (Dutch) traffic law does not state that you have to drive on the righthand side of the road but "to keep right as much as possible" as you are allowed to pass unexpected road blocks [38]). There are also catastrophic situations imaginable where the rules of traffic must be violated to avoid crashing.

To prepare for our driver's licence test, people typically learn from a textbook containing the rules of traffic, a few illustrations, and some explanations, which is much more efficient. One might therefore consider multi-modal AI-scenarios that would also involve learning from such textbooks. Even when many textbooks would be available, a pure (self-supervised) data-driven approach is likely to fail, too, as it would require learning to reason about traffic. But this is the same type of reasoning that proved to be problematic in the Alice and the boat-and-goat example.

Therefore, what is needed is a unified approach to AI that combines learning and reasoning, utilising both data and knowledge. We know the rules of traffic, so we should be able to use them to speed up the learning as well as to ensure that the learned model respects the rules of traffic.

2   Neurosymbolic AI for Learning and Reasoning

While there is a growing consensus that genuine AI should embrace learning as well as reasoning, there are different opinions as to how to integrate the two. To understand the issues, it is instructive to be more precise about machine learning and machine reasoning. The standard view on machine learning is that a machine learns when it gets better at a particular task with experience. Machine learning systems learn a function or model that maps inputs to outputs from data. Thus, they learn general patterns or functions from specific observations. According to Pedro Domingos [4], there are five different 'schools' or 'tribes' in machine learning that differ in the type of algorithm and model they use: the Bayesians (probabilistic graphical models), the Connectionists (neural nets), the Symbolistics (logic), the Analogizers (support vectors), and the Evolutionaries (genetic programs).

With regard to reasoning, different notions exist. At an abstract level, reasoning can be viewed as using more-or-less rational processes to infer new knowledge from one's existing knowledge (adapting [5]). What is important is that there is general knowledge as well as inference processes that allow one to draw conclusions about specific situations. This view on reasoning leaves room for two well-accepted types of reasoning in AI. The first is common sense reasoning [6], which is the human ability to use common sense knowledge to make presumptions about ordinary situations one encounters every day (adapting [7]). The other is the more formal type of rational reasoning that is guaranteed to be correct and that is found in probability theory and logic; it is this type of reasoning that corresponds to rational thinking according to Russell and Norvig [8]. Probability and logic also represent two of Domingos' five tribes mentioned above. This type of knowledge is traditionally understood as 'symbolic AI' and is incorporated in solvers that search for solutions to particular inference problems while providing guarantees [9]. As the models are symbolic, the knowledge is in principle interpretable and inference can be explained (through traces, derivations or proofs).

While it can be argued there has been a lot of progress in AI with large language models concerning common sense reasoning, it is hard to argue that such models are thinking rationally, let alone provide guarantees (see e.g. [10–12]). At the same time, the above examples of Alice and the boat-and-goat, show that there are still deficiencies, even with common sense reasoning.

Recent research highlights efforts to bridge the gap between LLMs and symbolic methods, attempting to imbue LLMs with more structured reasoning capabilities. Techniques such as "chain-of-thought prompting" [43] encourage LLMs to generate intermediate reasoning steps, loosely mimicking the systematic problem-solving processes seen in symbolic AI.

While more sophisticated large neural reasoning models such as OpenAI's o1/o3 and DeepSeek's R1/R2 achieve notable performance on symbolic-style tasks, they fundamentally differ from the aforementioned symbolic approaches that integrate explicit logical structure. In contrast, models like o3 and R2 operate as high-capacity sequence predictors without explicit symbolic grounding, treating reasoning tasks as monolithic pattern completion problems. As a result, while they can approximate logical behavior in distribution, they lack the modularity, verifiability, and systematic generalisation afforded by architectures that explicitly represent and manipulate symbolic knowledge. This divergence underscores a key limitation: without incorporating structured reasoning modules, large neural models risk conflating shallow statistical regularities with principled inference, limiting their reliability and interpretability in out-of-distribution settings.

A key advantage of symbolic reasoning systems over purely neural approaches lies in their ability to generalise universally across abstract structures without reliance on instance-level training. In relational logics, rules operate over variables and quantifiers, enabling inference about arbitrary sets of objects that satisfy certain conditions, independent of the particular instances seen before. For example, a rule like $\forall x.Student(x) \implies Mortal(x)$, stating that all students are mortal, generalises across the entire domain of discourse, enabling compact, compositional inference. In contrast, neural reasoning models such as o1/o3 and R1/R2 operate fundamentally on ground instances: they predict outputs for particular input examples seen during training, without an explicit representation of abstract variables or universal quantification. As a result, their generalisation is typically statistical rather than systematic, often failing when confronted with novel combinations or out-of-distribution configurations not reflected in the training data. This grounding limits their systematic generalisation capabilities, requiring explicit retraining or fine-tuning to adapt to new instances.

Moreover, symbolic systems are uniquely capable of reasoning about alternatives, counterfactuals, and hypothetical worlds. Logical formalisms allow the explicit representation of hypothetical or alternative worlds, making it possible to deduce what would follow under different assumptions. For instance, counterfactual reasoning—evaluating "what would happen if..." scenarios—is naturally handled by standard logical or modal extensions. Large Language Models, in contrast, are trained to approximate observed data distributions and typically lack intrinsic mechanisms to model counterfactual dependencies or maintain branching hypothetical states. Without explicit mechanisms for representing alternatives and dependencies, large neural models can only approximate counterfactual reasoning through interpolation between seen examples, lacking the rigor and transparency that symbolic systems offer. This fundamental asymmetry highlights why achieving deep generalisation and hypothetical reasoning remains an open challenge for neural systems, and why structured symbolic components are crucial for achieving robust, human-level intelligence.

The difference between learning and reasoning in AI is often linked to the System I and II characterisation of thinking that the Nobel prise-winner Daniel Kahnemann makes in his best-selling book "Thinking, fast and slow" [13]. System 1 is about fast thinking, it operates automatically and quickly, and it has learned patterns and associations, very much like what current neural networks and foundation models do. System 2, on the other hand, is about slow thinking, it demands complex multi-step computations or inferences, and it is deliberative and requires effort, which is reminiscent of reasoning.

The differences between learning and reasoning in AI can be summarised as follows. Learning using neural networks is data-driven, finds patterns and associations, produces black boxes which are hard to interpret, can be used for perception, operates at the subsymbolic level, and supports fast prediction. It involves inducing general patterns from specific observations. Reasoning on the other hand is knowledge-driven, based on declarative models (such as rules) that can be interpreted and explained, works at the symbol and knowledge level, and supports multi-step complex inferences. It involves making specific inferences using general laws. The subsymbolic level is hard to edit or intervene with, while the symbolic level does not account for perception directly.

The challenge and promise of learning and reasoning is to combine the best of both worlds. This challenge is pursued in the field of *neurosymbolic AI*, which is viewed as the 3rd wave in AI [14], has been named the 'most promising approach to a broad AI' [34], and is an innovation trigger according to Gartner [15]; it is prominently featured in DeepMind's latests results [16], and is enjoying a lot of attention. It has been the topic of AI debates, online summer schools attended by 1000 s of participants [18], and has given rise to a novel journal [19]. The goal of neurosymbolic AI is to answer the longstanding open question in AI about reconciling the symbolic and subsymbolic level, and the data- and knowledge-driven approach.

Neurosymbolic AI is based on some central ideas that can be cast into slogans. The first states that AI models should embrace knowledge as well as data:

$$\textbf{AIMODEL} = \textbf{KNOWLEDGE} + \textbf{DATA}$$

The use of the equation symbol reflects two important points. First, the sum of knowledge and data determines the quality of the model, and secondly, the use of addition indicates that knowledge can compensate for data: The more knowledge, the less data will be necessary and vice versa. This knowledge-data trade-off is analogous to the bias-variance trade-off in machine learning. This is well-illustrated by the ROAD-R dataset developed by Eleonara Giunchiglia and colleagues as a NeurIPS challenge [20] and illustrated in this demo [21]. The task in the dataset is to recognise objects in images about traffic, but unlike standard object recognition datasets, domain knowledge is specified as a set of logical rules. The idea is that the available knowledge should be used to constrain the learned models. Consider, for instance, the constraints that 'traffic lights do not move' and that 'the color of a traffic light is green or yellow or red'. When such knowledge is provided to a neurosymbolic AI system, the learned neural networks should respect these constraints.

Notice that the ensemble of knowledge and data might still be insufficient for learning a high-quality model. One indication of this is the presence of reasoning shortcuts [35], where the model appears to satisfy constraints but achieves this by misaligning its learned symbols or concepts with the designer's intended interpretation of the knowledge. Take the example of a car trained to stop for both pedestrians and red traffic lights. If these have always been observed together, the car might inadvertently map all pedestrians to traffic lights. This could lead to misbehavior in out-of-distribution scenarios. For instance, an ambulance can pass a red light, but it should still stop if there is a pedestrian. Concept alignment and grounding, which involves aligning a model's latent representations with human understanding, remains a critical open challenge in neurosymbolic AI and related fields.

The second slogan is that AI involves learners and reasoners.

$$\textbf{AI} = \textbf{LEARNERS} + \textbf{REASONERS}$$

While it is tempting to define reasoners as working on the symbol level only, this would deny the possibility that neural networks reason. Therefore, we will adapt Hector Geffner's terminology, who talks about model-free learners and model-based solvers. The learners are the neural networks operating at the subsymbolic level, while the solvers perform sound inference on a given symbolic model, where the semantics of the underlying formal model clearly define what sound means. Unlike model-free learners, model-based solvers provide guarantees w.r.t. the correctness of inference. This is essential for Trustworthy AI [23]; it is central to Technical Robustness and Safety (and its requirements towards accuracy, reliability, and reproducibility), and Transparancy (as inference in such models is traceable and therefore explainable).

From a Trustworthy AI perspective it is therefore relevant to distinguish between types of neurosymbolic AI systems based on whether the neural component or the symbolic solver is doing the reasoning, and on what type of interface and communication is used between the two components (see Kautz for an alternative conceptualisation [40]).

A. The symbolic solver as the reasoner. A standard, pipelining, technique in this category is to transform problems specified in natural language into a formal language that is accepted as input for a symbolic solver, and then mapping the result back. Prime examples that fall into this category include ChatGPT calling WolframAlpha to solve particular mathematical problems, and the many attempts of coupling LLMs to planners, e.g. [25]. The type of interface in the above two examples is very simple, it is more like a pipeline than a true interface.

Within category A, there also exist tightly integrated systems such as the neural concept learner and the deep probabilistic logics [26,27] that use deep interfaces. The interface for this type of systems is bi-directional, and differentiable, which allows to propagate the learning signal through both the symbolic and the subsymbolic components. Differentiability in this context can be interpreted in two ways. Strictly, it means each component must be continuous, smooth, and have derivatives at every interior point in its domain, enabling standard gradient-descent techniques. More broadly, it implies each component can compute a learning signal to improve the other, perhaps by synthesizing a training example as seen in abduction-based systems [45]. This leads to a tight and principled integration between the two models, and the solver and the neural network. This is the ultimate type of neurosymbolic AI system as it has the perception abilities of neural nets, the reasoning abilities of logic, and elegantly bridges the gap between the two. It is however also the most challenging type of system to develop and numerous research questions are still open.

B. The neural network as the reasoner. In this type of model, the subsymbolic model is restricted by the available knowledge. For instance, graph neural networks [28] are a prime example of this category as they can be trained to predict answers to queries that correspond to particular classes of logical or database queries. The underlying graph structure can be regarded as encoding the knowledge. A further example is that of techniques that use constraints to regularise (i.e. punish) the neural network when learning. Once the neural network has been learned, the constraints are no longer used for prediction or inference. It is therefore not guaranteed that the constraints are satisfied at prediction time. Important representatives of this stream of neurosymbolic AI systems include [29–31,39].

C. The symbolic solver as the reasoner with neural networks working in the same conceptual space. It is also interesting to consider a special type of class A problem in which the neural nets and the solver work together in the same conceptual space. For this variant, DeepMind's recent results in solving geometry problems are a prime example [16]. They use an LLM like model to guide a theorem prover towards finding a complex proof by proposing potentially interesting proof constructs. The LLM was trained from 100 million synthetic data example proofs. The conceptual space of the symbolic solver is thus the same as that of the neural network: proofs. The approach has been successful at competing in Math Olympiad problems [16].

Another example of this category is the SayCanPay planner [24], which trains three LLMs to select actions and to the check that the actions are feasible as well as effective. The LLMs are then used in a traditional heuristic search process to guide the search towards optimal plans. All three models as well as the plans are expressed in natural language.

Further extremes along the spectrum w.r.t. categories A and B are just a neural network (as in foundation or large language models), or just a symbolic solver, for instance, a planner. They either do not exploit knowledge or data, or consist only of either a subsymbolic or a symbolic component.

At the same time, it is also important to emphasise that when the distinction between data and knowledge becomes more blurred, the boundary between learning and reasoning becomes more fluid. For example, consider the natural language sentence "all students are mortal". Is it a data point (i.e., a sequence of tokens to be predicted) or a piece of knowledge (i.e., a logical rule to be formally interpreted)? Interestingly, regardless of the interpretation, both learning-oriented approaches and reasoning-oriented (iterative and structured) approaches are applicable [41]. For instance, in modern LLMs, training on that sequence using backpropagation or handling it through in-context learning represents two different strategies within this spectrum—one variational, the other iterative. Similarly, in neurosymbolic systems, reasoning over the corresponding formal rule $\forall x : Student(x) \rightarrow Mortal(x)$ can be framed either as variational inference (e.g., semantic-based regularisation) or as an iterative process (e.g., theorem proving).

3   Discussion

The above discussion and categorisation also points towards the interface as another, key differentiator among different neurosymbolic AI systems. Are the symbolic and subsymbolic components just loosely coupled in a pipeline, or is there a more fundamental deep integration that is bi-directional and differentiable, or are they just operating in the same conceptual space? Furthermore, is the solver or neural network just called once, as in the ChatGPT - WolframAlpha example, or is there an iterative process as in AlphaGeometry? The downside of loosely coupled neurosymbolic approaches is that the two components are independent of one another and that the interface is not really supporting learning. The challenge of deeply integrated neurosymbolic approaches is to design differentiable interfaces [42], differentiability that has proven to be extremely effective for machine learning and optimisation.

Traditionally, neurosymbolic AI focused on logic as the symbolic component. However, logic is discrete while neural networks are continuous. To bridge the two, one needs to introduce numeric information into the logics. This is the motivation for the view that the first author has been advocating [32] and that forms the basis for his research.

NeurosymbolicAI = NeuralNetworks + Logic + Probability

This also aligns with Russell and Norvig's view on rational thinking [8]. Probability, however, has to be interpreted broadly. Fuzzy logic has often been used as an alternative [30,31], it is typically more efficient, but suffers from semantic as well convergence issues in the optimisation [33].

The last slogan is important. It connects the stream on neurosymbolic AI to the five tribes of machine learning introduced by Pedro Domingos [4]. In his popular science book, he argues that the race is on to invent the ultimate learning algorithm, the quest for the ultimately master algorithm, which unifies the five tribes of machine learning: the Bayesians (probabilistic graphical models), the Connectionists (neural nets), the Symbolistics (logic), the Analogizers (support vectors), and the Evolutionaries (genetic programs). Analysing these different learning paradigms from a representational perspective reveals that they are essentially based on three representational paradigms: neural nets, logic and probabilistic graphical models as genetic programs and support vectors are typically not viewed as representational paradigms. Therefore, the quest for the master algorithm boils down to a quest for the master representation, the neurosymbolic representation that unifies Neural Networks + Logic + Probability and is to be used in neurosymbolic models. As a unifying representation, it should also have each of its constituents as special cases, that is, it should support the typical inference and learning processes that have made neural nets, logic and probability so successful for AI cf. [32]. It is the quest towards this master representation that the lab of the first author is pursuing [42], as many others in neurosymbolic AI.

As a final remark, AI is not just about rational thinking. Today, AI is more commonly viewed as rational acting, which is about taking the right actions and decisions. Selecting the right actions is based on rational thinking but also has an optimisation component: to decide what is the right action one needs to know what is the utility of different actions, and be able to optimise w.r.t. utility. This implies that AI, as well as learning and reasoning are intimately connected to optimisation. This connection is also clear from the many works lying at the intersection of constraint programming, operations research and machine learning (e.g. [36,37]). The ultimate challenge in this regard is to develop neurosymbolic reinforcement learning agents that combine perception with planning and learning [44]. After all, reinforcement learning captures the whole AI-problem in a nutshell.

Acknowledgments. The authors would like to acknowledge the TAILOR, a project from the EU Horizon 2020 research and innovation program under GA No 952215, and Trine Platou and Wannes Meert for interesting discussions and feedback, as well as Hendrik Blockeel, Jens Bürger, Vincent Derkinderen, Jaron Maene, Robin Manhaeve for comments on an earlier version of this manuscript.

Disclosure of Interests. The authors have no competing interests to declare that are relevant to the content of this article.

References

1. Nezhurina, M., Cipolina-Kun, L., Cherti, M., Jitsev, J.: Alice in Wonderland: Simple Tasks Showing Complete Reasoning Breakdown in State-of-the-Art Large Language Models. arXiv preprint arXiv:2406.02061 (2024)
2. Smith, G.: A Man, A Boat, and a Goat — and a Chatbot! Mind Matters (2024). https://mindmatters.ai/2024/05/a-man-a-boat-and-a-goat-and-a-chatbot/
3. Theorie-blokken.be. https://www.theorie-blokken.be/nl/gratis-proefexamen
4. Domingos, P.: The Master Algorithm: How the Quest for the Ultimate Learning Machine Will Remake Our World. Basic Books (2015)
5. Reason (2024). https://en.wikipedia.org/wiki/Reason
6. Davis, E., Marcus, G.: Commonsense reasoning and commonsense knowledge in artificial intelligence. Commun. ACM (2015)
7. Commonsense knowledge (artificial intelligence) (2024). https://en.wikipedia.org/wiki/Commonsense_knowledge_(artificial_intelligence)
8. Russell, S.J., Norvig, P.: Artificial Intelligence: A Modern Approach, 4th edn. Pearson (2016)
9. Geffner, H.: Model-free, model-based, and general intelligence. In: Proceedings of the Twenty-Seventh International Joint Conference on Artificial Intelligence (2018)
10. Kambhampati, S.: Can LLMs really reason and plan? Commun. ACM (2023)
11. Zhang, H., Li, L.H., Meng, T., Chang, K.W., Van den Broeck, G.: On the paradox of learning to reason from data. In: Proceedings of the Thirty-Second International Joint Conference on Artificial Intelligence, IJCAI-23 (2023)
12. Hazra, R., Venturato, G., Martires, P.Z.D., De Raedt, L.: Can Large Language Models Reason? A Characterization via 3-SAT. arXiv preprint arXiv:2408.07215 (2024)
13. Kahnemann, D.: Thinking, Fast, and Slow. Penguin Books (2012)
14. Garcez, A.D.A., Lamb, L.C.: Neurosymbolic AI: the 3rd wave. Artif. Intell. Rev. **56**(11), 12387–12406 (2023)
15. Gartner Research. Hype Cycle for Artificial Intelligence (2024). https://www.gartner.com/en/documents/5505695
16. Trinh, T.H., Wu, Y., Le, Q.V., et al.: Solving olympiad geometry without human demonstrations. Nature **625**, 476–482 (2024)

17. AI Debate between Yoshua Bengio and Gary Marcus. https://www.youtube.com/watch?v=EeqwFjqFvJA
18. IBM Neuro-Symbolic AI Workshop (2023). https://ibm.github.io/neuro-symbolic-ai/events/ns-workshop2023/
19. Neurosymbolic Artificial Intelligence. IOS Press. https://journals.sagepub.com/home/nai
20. Giunchiglia, E., Stoian, M.C., Khan, S., Cuzzolin, F., Lukasiewicz, T.: ROAD-R: the autonomous driving dataset with logical requirements. Mach. Learn. **112**(9), 3261–3291 (2023)
21. https://dtai.cs.kuleuven.be/projects/nesy/roadr.html
22. Manhaeve, R., Dumancic, S., Kimmig, A., Demeester, T., De Raedt, L.: Deepproblog: neural probabilistic logic programming. In: Advances in Neural Information Processing Systems, vol. 31 (2018)
23. European Commission. Ethics guidelines for trustworthy AI (2019). https://digital-strategy.ec.europa.eu/en/library/ethics-guidelines-trustworthy-ai
24. Hazra, R., Dos Martires, P.Z., De Raedt, L.: Saycanpay: heuristic planning with large language models using learnable domain knowledge. In: Proceedings of the AAAI Conference on Artificial Intelligence, vol. 38, no. 18, pp. 20123–20133 (2024)
25. Silver, T., Dan, S., Srinivas, K., Tenenbaum, J.B., Kaelbling, L., Katz, M.: Generalized planning in PDDL domains with pretrained large language models. In: Proceedings of the AAAI Conference on Artificial Intelligence, vol. 38, no. 18, pp. 20256–20264 (2024)
26. Marra, G., Dumancic, S., Manhaeve, R., De Raedt, L.: From statistical relational to neurosymbolic artificial intelligence: a survey. Artif. Intell. **328**, 104062 (2024)
27. Manhaeve, R., Dumancic, S., Kimmig, A., Demeester, T., De Raedt, L.: Neural probabilistic logic programming in DeepProbLog. Artif. Intell. **298**, 103504 (2021)
28. Graph Neural Network (2024). https://en.wikipedia.org/wiki/Graph_neural_network
29. Jingyi, X., Zhang, Z., Friedman, T., Liang, Y., Van den Broeck, G.: A semantic loss function for deep learning with symbolic knowledge. In: ICML (2018)
30. Badreddine, S., d'Avila Garcez, A.S., Serafini, L., Spranger, M.: Logic tensor networks. Artif. Intell. **303**, 103649 (2022)
31. Diligenti, M., Gori, M., Saccà, C.: Semantic-based regularization for learning and inference. Artif. Intell. **244**, 143–165 (2017)
32. Manhaeve, R., Marra, G., Demeester, T., Dumančić, S., Kimmig, A., De Raedt, L.: Neuro-symbolic AI = neural + logical + probabilistic AI. Front. Artif. Intell. Appl. (2021)
33. van Krieken, E., Acar, E., van Harmelen, F.: Analyzing differentiable fuzzy logic operators. Artif. Intell. **302**, 103602 (2022)
34. Hochreiter, S.: Toward a broad AI. Commun. ACM **65**(4), 56–57 (2022)
35. Marconato, E., Teso, S., Passerini, A.: Neuro-symbolic reasoning shortcuts: mitigation strategies and their limitations. arXiv preprint arXiv:2303.12578 (2023)
36. Mandi, J., Demirovic, E., Stuckey, P.J., Guns, T.: Smart predict-and-optimize for hard combinatorial optimization problems. In: Proceedings of AAAI (2020)
37. Lombardi, M., Milano, M., Bartolini, A.: Empirical decision model learning. Art. Intell. (2017)
38. Prakken, H.: On the problem of making autonomous vehicles conform to traffic law. Art. Intell. Law(2017)
39. Van Baelen, Q., Karsmakers, P.: Constraint guided gradient descent: training with inequality constraints with applications in regression and semantic segmentation. Neurocomputing **556** (2023)

40. Kautz, H.: The third AI summer: AAAI Robert S. Engelmore memorial lecture. AI Mag. **43**(1), 93–104 (2022)
41. Marra, G.: Bridging symbolic and subsymbolic reasoning with MiniMax Entropy models. PhD. thesis (2020)
42. De Raedt, L.: ERC AdG DeepLog (2023). https://wms.cs.kuleuven.be/cs/onderzoek/deeplog
43. Wei, J., et al.: Chain of thought prompting elicits reasoning in large language models. In: NeurIPS 2022 (2022)
44. Shindo, H., Delfosse, Q., Dhami, D.S., Kersting, K.: BlendRL: a framework for merging symbolic and neural policy learning. In: Proceedings of the International Conference on Representation Learning (ICLR) (2025)
45. Hu, W.C., Dai, W.Z., Jiang, Y., Zhou, Z.H.: Efficient rectification of neuro-symbolic reasoning inconsistencies by abductive reflection. In: Proceedings of the AAAI Conference on Artificial Intelligence, vol. 39, no. 16, pp. 17333–17341 (2025)

Resource-Aware Machine Learning

Katharina Morik[✉][ID]

Lamarr Institute for Machine Learning and AI, Dortmund, Germany
katharina.morik@tu-dortmund.de
https://lamarr-institute.org

Abstract. Resource-aware machine learning has two main motivations. On the one hand, the internet of things, embedded systems, edge AI, and federated learning demand machine learning to manage computation with less resources, i.e., runtime, memory, communication, and energy. On the other hand, learning large models need to become more aware of resources, because they consume too much. Regarding the climate change, saving resource consumption has become an urgent need. Both motivations lead to the same scientific subject, namely the design and implementation of machine learning algorithms that are optimized to get along with less resources than a straight-forward version. Where embedded systems always dealt with various computing architectures, the larger models and finally the large language models rely on efficient chips with parallel processing. In any case, the implementation on a certain hardware needs to be taken into account. Given the huge environmental impact of computing, the choice of an implemented model should now be based on how low its resource consumption is. Hence, it is important to measure, test, and report model features such that users can easily compare the implemented models and choose the one with a minimal footprint.

This chapter introduces the facets of resource-aware machine learning indicating references to literature that offer in-depth studies. After a recap of sustainability demands, the resource consumption and approaches to reducing it are shown. Since new results are published every day, this chapter cannot even attempt to provide a survey of the plethora of papers on energy-saving models. It structures the field and carefully selects relevant literature that eases to catch up with new models. A method for testing and reporting is proposed that visualizes the results in analogy to care labels or electronic property cards. The hope is that people select models with minimal resource consumption and this, in turn, motivates the developers to bring resource-efficient systems to the market.

Keywords: Machine Learning for Sustainability · Resource-Aware Machine Learning · Energy Consumption of Machine Learning

© The Author(s), under exclusive license to Springer Nature Switzerland AG 2026
M. van Leeuwen and J. Vreeken (Eds.): Arno Siebes Festschrift, LNCS 16067, pp. 109–126, 2026.
https://doi.org/10.1007/978-3-032-03028-3_7

1   Introduction

Artificial intelligence (AI) and statistics started with the analysis of some short tables of observations. Relational databases with their several linked tables became the standard in the database field. The next step moved from Very Large Databases (VLDB) to Big Data with the characteristics of volume, velocity, variety, and veracity. 'Volume' characterizes the huge amount of stored data, where 'velocity' refers to streaming data. Both types challenge the algorithms for the storage, access, and summary of huge masses of data. Hence, compressing the data became the hot topic and Arno Siebes and his group were heading the field [55]. The KRIMP system compressed stored itemsets [61] as well as streaming ones [33]. This integration of Knowledge Discovery in Data (KDD) and Machine Learning (ML) has inspired many further studies. Now, with the very large AI models, we face again the need to reduce the resource demands. This chapter advocates resource-awareness for AI models. Usually, streaming data and Internet of Things (IoT) approaches on the one hand and Large Language Models (LLMs) on the other hand are discussed separately in different communities. Here, however, we follow the Siebes group to inspect stored and streaming data, together. The communities should know about each other and read the literature, because analogies from one field to the other might be inspiring[1].

2   Sustainability Demands

Let us start with a short recap of the global initiatives for sustainability. Since the famous book "The Limits to Growth" of the Club of Rome in 1972 [39], sciences have investigated feedback loops and emerging behaviors that determine how long environmental processes can support population and economic growth of mankind. In 1987, when being Norwegian Prime Minister, Gro Harlem Brundtland defined the term 'sustainability' for the United Nations. In 2000, the UN declared millenium development goals and the sustainable development goals 2012. The Conference on Climate Change achieved the Kyoto protocol in 1997 and the Paris agreement in 2015. Research has been organized in the Intergovernmental Panel on Climate Change (IPCC), which received together with Al Gore the Nobel peace price in 2007. The most recent assessment report is from 2023 and the 2027 IPCC report has been already planned. All these activities are to strengthen our response to the threat of climate change and to increase our efforts to eradicate poverty.

Focusing on Greenhouse Gas (GHG) has become the key indicator of global warming. Carbon Dioxide (CO_2) reached 410 parts per million in 2019 for the

---

[1] I am well aware that this chapter cannot be as inspiring as the lectures or texts of Arno Siebes. It illustrates my sometimes annoying habit of recommending literature to authors who proudly present the result of some months of work and not (yet) appreciate further reading. In a book chapter, though, references cannot be misunderstood as criticism and, hence, might be more welcome.

first time in the last 2 million years. Human-caused GHG leads to $+1\,°C$ or $+2\,°C$ warming. In 2023, we have $1.1\,°C$ warming already. This is dangerous for human health. Heat and relative humidity poses a risk of mortality [29]. Immediate actions towards saving resources are necessary, particularly regarding energy efficiency, water management, and materials efficiency (recycling and reuse)[2]. The first answers to these challenges on behalf of ML were directed towards applications of learned models that decrease resource consumption in traffic, production, agriculture, medicine, geoengineering, and biodiversity [32,41,53]. Some already then inspected the energy efficiency of computing and computing architectures themselves, e.g., the advantages of ARM CPUs and scalable storage solutions for less power consumption [20]. In contrast to ML helping the environment, there are high environmental costs of ML and computing, in general. We summarize these for computing centers in the next sections, before we look at Deep Learning and Large Language Models (LLMs) and then embedded systems and the Internet of Things (IoT).

2.1 Computing Centers

The very basis of computer resource consumption is the computing center, be it used locally or as a cloud. Hence, characteristics of a system's carbon footprint need to take into account that of the computing center on which it is executed.

Water. Computing centers need cooling. First, the placement of the computing center influences the amount of necessary cooling. Preferred are locations in the rocks like the center of the internet provider *Bahnhof* in Stockholm or, generally, areas with an average low air temperature like, e.g., in Canada, Finnland, Sweden, or Norway. Second, cooling towers consume some water due to evaporation (water consumption), but most of it can be reused several times. However, mineral or salt buildup demands to add fresh clean water, continuously. Water withdrawal refers to added water, no matter whether this is necessary because of the water quality or because of evaporation. The environmental reports of Google and Microsoft include many projects improving water management. Google states that 22% of the total data-center water withdrawal was reclaimed wastewater or other non-potable water and one-third of the data-centers use air-cooling[3] Microsoft claims that new data-centers will consume zero water for cooling and rainwater replaces some water withdrawal[4] Water Usage Effectiveness (WUE) is the ratio of water consumption to server energy consumption. For *GPT-3* training, overall 1287 MWh have been used and "...training

---

[2] Disaster risk management, human migration and relocation, resettlement are important actions for the survival in times of climate change. Reactions to floods and storms need to be prepared, if these disasters cannot be prevented. The northern part of the globe and, in general, higher areas are less affected. Movements from destroyed areas to still livable ones need to be organized.

[3] https://sustainability.google/reports/google-2024-environmental-report/.

[4] https://www.microsoft.com/en-us/corporate-responsibility/sustainability/report.

the *GPT-3* language model in Microsoft's state-of-the-art U.S. data-centers can directly evaporate 700,000 liters of clean freshwater, but such information has been kept a secret." say Peifeng Li and colleagues [34]. Hence, water footprint should be reported along with energy footprint.

Energy and CO2. Computing capacity is increasing. A study for Germany revealed that the workloads per kWh increased by 500% from 2010 to 2022. Due to efficiency, the energy consumption increased only by 70% at the same time, reaching 17.9 billions kWh per year. Given the average German mixture of energy production, this corresponds to 7.8 million tons of CO2 [24]. This is not intended to play the cost to the environment down, but it shows that efficiency really makes a difference. On the other hand, the efficiency is not saving us. In contrast, rebound effects may even lead to more generous energy use [35].

A study by the Project *Bytes2Heat* for the German Federal Ministry for Economic Affairs and Climate Action lists 98 use cases of waste heat utilization in agriculture, industry, and heating of buildings, worldwide[5]. For instance, in Braunschweig, from the data centre as a heat source, 250 kW of the waste heat with a temperature of 18–25 °C is decoupled and used via a 3 km heat network by the neighboring residential quarter with 400 houses.

The decrease of environmental cost due to efficiency and due to reuse should not be mixed up with the so-called Renewable Energy Certificates (Recs). These are just bought and do not necessarily relate to the particular computer center, but can be anywhere. Even already existing trees can be claimed as decreasing CO2 emissions. Recs refer to the overall company that aims at being labeled climate neutral. TikTok, for instance, partnered with Climeworks to remove 5100 tonnes of CO2 [6] However, the true important figure is the location-based emission of CO2 and whether additional actions are undertaken. Only this should be reported for the computing centers.

Finally, the production of the chips and the overall computing devices including batteries is also an important environmental cost factor.

2.2 Deep Learning and Large Language Models

The first publications that pointed at the energy demand of deep learning were the ones on natural language processing [57] and on a carbon emission calculator for neural networks [31].

Facebook used 2,638,000 kWh of electricity and emitted 1,000 tons of CO2 when developing their new *Llama* models. To train the 4 different sizes of *Llama* they used 2048 Nvidia A100 – 80 GB GPUs for a period of approximately 5 months. Producing that amount of electricity is estimated to have led to the emission of 1,015 t CO2e [60]. The energy consumption from developing the *Llama* models cover the development of all four models including failed runs,

---

[5] The consortium consists of DENEFF, empact, and University Stuttgart with its Institute of Energy Economics and Rational Energy Use bytes2heat.de.

[6] https://tinyurl.com/27lhqkjh.

where the estimate of the electricity consumption from training *GPT-3* probably only considers the training of the final model, i.e., 14.8 days of training instead of 5 months [36].

Large Language Model (LLM). LLMs even exceed the deep learning demand of computing capacity. Theoretical analyses and empirical evidence consistently demonstrate that scaling up model size leads to substantial performance improvements [4]. However, training such large-scale models poses considerable challenges, demanding extensive time, energy, and financial resources.

Reasoning Models. The reasoning models based on *chatGPT*, i.e., OpenAI's *o3* and *o4* spend more compute time for one user question than regular chats, which increases the used energy and water consumption, considerably. If we assume that there are 500 millions of GPT users and at least 10% of them ask for help with some, let's say, math or chemistry problem, we have already 50 million questions with a possibly 100 times higher resource demand. Moreover, in March 2025, with the easy to use image generation of OpenAI's image creator *GPT4o*, the creation of images in, e.g., Lego or Studio Ghibly style has exploded. The number of users, the rollout, is enormous for easy to ask systems! Assuming that each creation uses around 2 liters of water, we reach the incredible amount of around 1 billion liters used just for funny images. These examples highlight the need for more efficient training and inference methods.

2.3   Embedded Systems and Internet of Things

Many devices are becoming smart, today. They perceive their surrounding and send measurements to cloud servers or directly to a certain engine. In 2024, there were 18.18 billion connected IoT devices, worldwide[7]. This number covers connected vehicles, payment terminals, remote diagnostics, consumer internet and media devices, remote process control, and many more. Given that only a few of the electrical and electronic equipment is recycled or reused, this is wasting resources: in 2020, in Europe alone, there were 4.7 million tons of scrap according to the Federal Statistical Office of Germany. Alone 1,339.51 smartphones have been sold in 2023[8] Regarding the carbon footprint of smartphones, the production dominates that of usage, by far. In particular, smartphones are not yet recycled so that the energy, water consumption, and the materials are just wasted. Moreover, given the huge number of smartphones, also their usage energy consumption should not be forgotten.

Learning on Small Devices. In 2011, the Collaborative Research Center 876 started with the aim to reduce memory, communication, and energy needs of

---

[7] https://www.statista.com/statistics/1194701/iot-connected-devices-use-case/.

[8] https://www.statista.com/statistics/263437/global-smartphone-sales-to-end-users-since-2007/.

ML. It investigated and invented machine learning methods for small devices and stressed the importance of real-time inference and communication with smallest possible latency [44]. Although large models are currently receiving spectacular attention, learning on small devices and real-time predictions actually have a huge impact on, e.g., fraud detection, autonomous driving, health monitoring, and trading. Embedded devices in warehouses and factories help optimizing the logistic and production processes. Personal watches are used for self-optimization, but also for fall detection of elderly people. Recommendation systems that are based on current user activities are superior to those that refer to long ago interests. The company Apple clearly outlines the importance of ML on small devices and makes it one important pillar of their business [25].

2.4   How to Measure CO2 Consumption

Energy consumption can be measured at several levels, ranging from the circuit over the processor until the operation system and finally an application. Energy is the integral over the power required per operation.

Low-Level Energy Accounting. Measuring voltages and currents on a hardware at hand is relatively easy. Without access to the particular hardware, power and energy models need to be designed and carefully validated [37]. For wireless sensors used by IoT applications, energy models need to be integrated into the low-level code. An example of energy accounting for in-house warehousing is detailed in [38]. Their online model of a device component is a Finite State Machine that annotates each state by average power costs and transitions by average energy costs. Counting the transitions and multiplying the state costs by how long the machine is in this state gives the energy accounting. This is then integrated into the compiler.

Ultra-Low Power Devices. If we go further to batteryless sensors that harvest energy from photovoltaic, the energy management is even more challenging. The sensor only works, if there is some light [56]. The MiroCard, for instance, is a batteryless smart-card that is activated when the user exposes it to light [18]. It hosts an Internet communication protocol and several sensors, e.g., an accelerometer. Even such Zero-Power devices can gather data – here, gestures of users holding the MiroCard – transmit them to a Raspberry Pi for training, and apply the learned model to classifying gestures of users. A case study with the proprietary light-weight neural net *Lt-Wt* shows that the classification takes around 27 ms after the boot-up, and the *Lt-Wt* needs between 20 and 35 milliwatt (mW) [19]. Hence, even on severely restricted devices, ML is possible using resource restricted algorithms (see also below for a learning algorithm on ultra low-power devices).

Programming Languages. Already programming languages have their energy consumption profiles. A case study on 10 problems which have to be solved by the same algorithms compares the implementations in 27 different languages using standard compilers. C is best in using minimal time, memory, and energy, followed closely by Rust and C++. Python is almost the worst, only Perl is even worse. For Python, time and energy, and also memory and energy strongly correlate [46]. For embedded systems, there is a study on programming language's energy-delay. It shows that C is the best in arithmetic, compression and concurrency, while C++, Go, and Rust are the runners-up. R, Perl, Swift, and Java show weak performance [17].

Computing Architectures. Different hardware architectures consume more or less energy. In general, Field Programmable Gate Arrays (FPGAs) use the least Watt, CPUs come next, followed by GPUs and finally, the Quantum Annealer consumes most. Measurements supporting this ranking have been set up for Quadratic Unconstrained Binary Optimization (QUBO) and for the deployment of learned decision trees [42]. Over several data sets and parameter settings the QUBO experiment showed the following order of magnitude:

- Field Programmable Gate Arrays (FPGAs), 10^0W,
- CPU (Intel Core i7-9700K)10^1W,
- GPU (Nvidia GEFORCE RTX 2080 Ti)10^2W,
- QA (Quantum Annealer, IBM)10^4W.

Another study compares different implementations of applying a learned Decision Tree (DT) model [6]. Energy consumption is measured for FPGA (Xilinx Artix-7 Z-7020 FPGA with 53200 lookup tables, 106400 flip-flops (FF) in total combined with 4.9 Mb block ram and 220 DSP units) and an ARM processor (Cortex- A9 with 666 Mhz, 512 Mb DDR RAM and 512 Kb cache). Each learned tree contains an average of 1349 nodes and roughly 675 different paths from the root node to a leaf node. Throughput is measured as elements per millisecond, energy consumption as nano Joule per element. The native implementation on FPGA uses 0.008W per element to classify, on the ARM processor 1.53 W per element to classify [42]. Also the memory architecture plays a critical role in modern computing architectures. Approximate or non-volatile memories decrease the energy needs while at the same time increase the bit error rate. Specialized tensor processors also reduce the carbon footprint. Due to rapid developments, the computer architecture race is still open.

Available Measuring Software. For hardware components and the code that runs on them, some companies have introduced fine-grained energy models. With Intel's Sandy Bridge microarchitecture comes along the *Running Average Power Limit (RAPL)*. *RAPL* is also capable of measuring the energy consumption of whole CPUs, processor cores, memory controllers, even at kernel level. A comparison with a manually instrumented board showed some deviations of the *RAPL*, but an enhanced framework for energy measurements has been developed [21].

The in-depth paper [22] on how to report carbon footprints offers an *experiment-impact-tracker* that should be easy to deploy for ML papers to report carbon impact summaries. It demands that all ML experiments should come along with a report of the carbon footprint. Specialized proposals are the carbon emission calculator for neural networks [31] and the tools for estimating the energy consumption [58]. A careful comparison of diverse large language models by David Patterson et alii [45] shows how to calculate greenhouse emissions of deep learning. He states that *GPT-3* used 552.1 tons of CO2. Since Hugging Face is a large repository, the investigation of the carbon footprint of the ML models, which are stored there, is extremely useful [8].

The *LLMCarbon* tool starts even before training a model. Beyond *mlco2*[9], it predicts the carbon footprint of a new model based on the architectural description (e.g., number of parameters, number of training tokens, and the parallelism setting) and the hardware units (CPU, GPU, TPU, drivers, memories) plus the power usage effectiveness of the respective data-center. It uses the *experiment-impact-tracker* but adds prediction models for the architecture and hardware units. Moreover, it predicts the carbon footprint for training, inference, and cloud storage as well for regular as for mixture-of-experts models [13].

Predicting upcoming energy consumption is also and perhaps even more important for IoT systems. Learning a generative probability distribution over all the device's states allows to model the behavior of an IoT device and a 5G network under various workloads. Hence, the networked device deduces its energy model from its own resource usage data. An implementation of this approach puts Markov Random Fields (MRF) working with integer numbers to good use in resource-aware 5G and IoT systems [50].

3   Reducing Resource Consumption of Machine Learning

3.1   Less Hungry Deep Learning and Large Language Models

More efficient deep learning has been achieved by quantization, as low-bit arithmetic kernels can save memory and accelerate computations when used appropriately. Techniques from lossy compression may inspire the minimization of memory while keeping the accuracy high enough [12].

Binarized Neural Networks. Binarized neural networks have been introduced in 2016 [28]. Training the network with binary weights and activations reduced the memory consumption considerably. Important is the use of POP-COUNT (XNOR) for the most used operation $\sum_i f_i w_i$. This makes the forward pass extremely efficient and the binarized neural network inference 7 times faster. An example of quantizing trained neural networks is the *Fast Inference* system which maps the learned weights from 32bit floating point numbers to fixed point numbers and then to binarized $+1, -1$. This allows to use the processor's POP-COUNT so that the inference uses the least energy [7]. The *Fast Inference* was

---

[9] https://mlco2.github.io/impact is an easy to use carbon estimator.

necessary for realtime classification of a stream of astrophysical particles, 60 events per second, each 3 MB of raw data. The energy-efficient storage for low energy neural networks on the edge is responsible for further savings of energy consumption.

Quantized LLMs. For LLMs, the low bit quantized *LLama3* has pushed the post-training quantization to 1–8 bits and exploits diverse quantization methods for LLM post-training [27]. The *BiLLM* method selects salient weights of a pretrained *LLama* model and minimizes the compression loss through a binary residual strategy. The result is a model with only 1.08 bit weights. For the energy consumption, it is important that *BiLLM*'s binarization only takes about 0.5 h on a single GPU [26].

Quantization of training a LLM has also been proposed [62]. They reduce full precision floating point 32 or 16 to 4-bit. A quantization estimator corrects the error of the gradient updates. The activations are corrected by detecting and correcting outliers which are often observed during training. The distribution of activations is the basis for the outlier detection.

An important part of LLMs is the tokenization [1]. The open-source model for all European languages, *Teuken7B*, has shown that multi-lingual tokenization for all the languages saves energy in training as compared to training for each language independently[10]. In contrast, Aleph Alpha and colleagues have trained tokenizer-free, i.e., just on n-grams [11]. Also their model used considerably less energy than Meta's *Llama3*.

The fast development of less energy-hungry training and inference models should not let us become too optimistic, though. The rebound effect could well become true through more and more capabilities and more usage of LLMs [35].

DeepSeek and OpenHands Agents. OpenAI has enhanced LLMs by reasoning skills, presenting *GPTo1* (later replaced by o3) and *GPTo4*. In January 2025, a distillation of reasoning patterns has been published: *DeepSeek* [10]. The use of Reinforcement Learning (RL) on top of a base model leads to excellent performance on the reasoning benchmarks[11]. The sampling for self-evolution of the model and its inner model-of-experts approach brought upon a stunning performance. The distillation into smaller models results in saving energy resources. Since *DeepSeek* is open source, their RL framework could be exploited by the developers of other open source LLMs, as well.

In March 2025, an agent for software development has been uploaded to HuggingFace, *OpenHands*[12]. It is based on the open-source LLM *Qwen2.5* of Alibaba [63]. Its performance exceeds that of the larger *DeepSeekR1 V3*, only *DeepSeek*

---

[10] https://www.iais.fraunhofer.de/de/geschaeftsfelder/speech-technologies/conversational-ai/opengpt-x.html.

[11] https://huggingface.co/deepseek-ai/DeepSeek-R1.

[12] https://www.all-hands.dev/blog/introducing-openhands-lm-32b----a-strong-open-coding-agent-model.

V3 0324 is slightly better. Open source LLMs including those that generate images, solve problems, or code software are now published in rapid succession. Any list of models would be outdated by the publication date of this chapter. Hence, we only want to stress that the novel models demand less resources and hope that this trend continues, but note, that the roll-out is enormous.

Some LLMs are already small enough to be used on Raspberry Pi so that they are ready for edge computing. There, real-time behavior is required and the bounds for acceptable latency and model size are low. The HuggingFace leaderboard for Edge LLMs[13] shows that *DeepSeek V2 Lite* (10.36 GB) is ranked second, the first is Google's *gemma-2-9b* (10.796 GB). The smaller *DeepSeek V2 Lite* with 8.901 GB is 7th.

In the next Sect. 3.2, learning on the edge will be discussed regarding the resource consumption for other learning methods.

3.2   Compressing Models for the Internet of Things

The large number of embedded devices, communicating sensors, Apps on diverse devices has been said to communicate around 201,010 GB per month[14] with a power consumption of around 0.3 kWh/GB [51], resulting in a total energy consumption of 6 TWh per month. Where these figures are from 2018 and 2019, we may well assume that the amount has not decreased.

Integer MRFs. In Sect. 2.4, ultra-low power devices have already been introduced. Surprisingly, models with high expressiveness like Markov Random Fields (MRF) can be learned on restricted processors, even those that lack floating point arithmetics. Floating point arithmetics costs more clock cycles than integer arithmetics. Less cycles means conservation of power. Now, Nico Piatkowski's idea was to restrict the parameter space of MRFs to integer numbers [48][15]. For undirected models like MRFs, the parameters can be mapped to the integer domain and then approximations are computed for the marginal probabilities, the Maximum A Posteriori probability and the Maximum Likelihood Estimate – all only using integer values. Of course, the integer parameter space requires a novel method for belief propagation. Based on the observation that the magnitude of messages from one node to the other actually corresponds to the probability, a novel message passing algorithm has been developed. It approximates the integer bit length. The approximation is not lossless, but a bound of the Kullback-Leibler divergence between the true probability mass function and the approximated one has been proven [47]. Indeed, the number of clock cycles for training is reduced and the processing speeds up which has been shown on both, an ARM11 and an Intel architecture.

Inference on graphical models includes the handling of integrals. Where stochastic quadrature in general is not tractable, the complexity of a novel par-

---

[13] https://huggingface.co/spaces/nyunai/edge-llm-leaderboard.

[14] https://datareportal.com/reports/digital-2019-global-digital-overview.

[15] This does not refer to the state space of the model.

allel quadrature algorithm is independent of the dimension and, hence, much faster, which saves energy [49]. Whether the recently proposed probabilistic graphical circuits [16] are also reducing energy consumption has not yet been published, but more is to come.

Decision Trees and Random Forests. IoT applications often handle streaming data in real-time. The observations are received by a resource-restricted device and need to be filtered directly. Filtering by decision trees or random forests on resource-restricted devices is challenging, because of their large memory footprint and intensive memory accesses. Sebastian Buschjäger has investigated ways of implementing random forests on different computing platforms. The basic representation for the analysis is a probabilistic execution model for binary decision trees. It represents the probabilities of an instance taking a certain path in the tree as a series of Bernoulli experiments. At each node i, the probability that the path continues with the left child j and the probability that it continues with the right child k can be estimated by counting the training instances along the paths [6]. There are mainly two ways to implement decision trees, the naive implementation in terms of binary search trees and the if-else-trees which unroll the tree and reduce memory accesses. Based on the probabilistic model, the resource demands of both ways were theoretically analyzed for CPU and for FPGA. A code generator implements a learned classifier on a given computing architecture so that empirical experiments complete the study. FPGAs use the least energy per element for all implementations. Where this study leaves the learned decision tree as it is, leaf refinement decreases the expected number of comparison and memory accesses [52].

Integrating leaf refinement with ensemble pruning, *LRL1* jointly prunes and refines trees through L1 regularization [5]. By clustering the data, representative trees for each cluster are selected. A further criterion is the diversity of the members of the resulting ensemble. The objective function of Stochastic Gradient Descent (SGD) optimization combines leaf refinement and pruning together with a regularization that prioritize smaller ensembles. An even further compression for edge devices, Splitting Stumps Forests, *SSF* selects nodes based on the probabilistic model and turns them into small balanced trees of depth 1, breaking the path into smallest if-else trees, thus forming an ensemble of stumps. An example receives a value for each stump, hence a binary vector for the ensemble. Logistic regression is then trained on the transformed examples [2]. Regarding small memory budgets, *SSF* outperforms *LRL1* in terms of accuracy.

Distributed Learning. The communication of devices in the IoT world consumes considerable amounts of energy. If ML is performed directly at the sensing device, communication is reduced. Federated learning is well suited to learn a global model from several local models with less communication. Moreover, distributed learning preserves privacy. An example is the recent work on binary matrix factorization used for federate learning through a proximal gradient descent algorithm [9].

Dynamic averaging of data streams on resource-constrained systems communicates local summaries only periodically or dynamically, i.e. only if something happens that differs from the global mean. A particular energy saving method of dynamic averaging deploys only integer computations [23]. An estimate of the energy consumption of the centralized and of the distribute approach shows that the centralized approach needs about 67 times more energy than the parallel one. Even large amounts of local learners do not change the picture.

Taking distributed machine learning to the extreme, we might even drop the constraint that the learned model is continuously improved and redeployed on a virtual machine. Instead, functions of a commercial cloud are just computed on demand and the life-span of the respective container ends after answering the request. Also this is saving energy! However, serverless training requires communication and synchronization of data between the worker nodes (i.e., the serverless functions employed for the training process) and the orchestrator (e.g., parameter server)[16]. A first approach for serverless Random Forests variants has recently been introduced with the middleware *STRATA* [59]. It shows that *STRATA* needs less data and less training time to reach about the same accuracy as with a centralized procedure.

4   How to Report Measurements to Users

If we want users to choose resource-aware implementations of ML, the resources need to be reported together with other properties that are important for the model selection. Some colleagues proposed sets of properties for certification, namely the *FactSheets* [3], the *Model Cards* [40], and some others [22,54]. They aimed at comprehensibility and wrote texts that explain the properties of the model. They all did not include theoretical results proving properties of learned models and did not test whether the implemented model actually shows the properties that are proven for the learning method. Most of them did not take into account the particular hardware on which the model was to be executed.

Labeling Implemented ML Models. We proposed the *Care Labels* which characterize features of a learned model that is implemented on a certain hardware [43]. On the basis of a proof repository and profiling data sets, the certification suite generates large sets of experiments that run the implemented method on a given hardware and report the performance for each property. The performances are then ranked and labeled as 'A, B, C' according to their fulfillment. Some static properties like 'generative vs. discriminative', 'outputs uncertainty', 'can be tested for robustness', theoretical bounds of time and memory consumption, 'might be run on ultra-low power devices' and 'can be used with streams' are indicated, as well. The runtime, memory use and energy consumption is monitored by the tests. First, diverse implementations of MRFs, then those of

---

[16] A serverless environment is, e.g., Mesosphere Marathon https://mesosphere.github.io/marathon/.

Neural Networks have been studied. The method has been pushed further such that user interaction is possible [14]. The Sustainable and Trustworthy Reporting tool *STREP* offers a library for comparing various model properties and generates labels on the basis of the evaluations. It covers diverse ML methods, not just LLMs. The efficiency of ImageNet models, of those from the RobustBench leaderboard, those from the Monash forecasting time series and some from the Papers with Code collection have been tested and characterized by diverse visualizations. The trade-off between predictive performance and resource efficiency is shown by the various model evaluations. Index scaling delivers the rating for the care label visualization (see Fig. 1). The visualization has been found helpful by practitioners from different fields, backgrounds, and levels of AI experience [15].

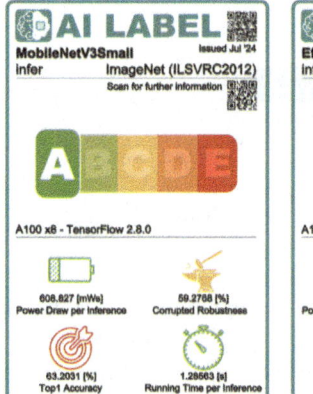 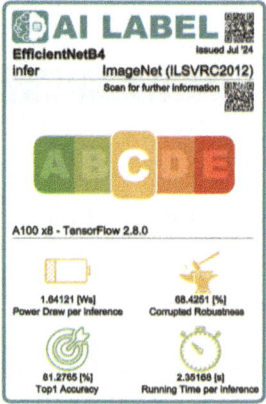

Fig. 1. CareLabel generated by STREP

Labeling LLMs. With the extraordinary opportunities of HuggingFace, a software library for energy scoring of LLMs has been published together with blog posts and a leaderboard[17]. The hardware setting is fixed for all models. Single consumer GPU, single cloud GPU, and multiple cloud GPU are the given choices. Energy reduction by using less consuming computing architectures is not taken into account. This can be explained by the focus on LLMs for the tasks of

---

[17] Sasha Luccioni, Margaret Mitchell, Yacine Jernite, Regis Perrard are named contributors from HuggingFace, in addition Sara Hooker from Cohere, Carole-Jean Wu from Meta, Emma Strubell from Carnegie Mellon University. Thanks go to the AI Convergence Challenge organized by the French Government as part of the 2025 AI Action Summit in Paris, France. https://huggingface.co/spaces/AIEnergyScore/Leaderboard.

text generation, image generation, text classification, image classification, image captioning, and summarization. The visualization can be seen in Fig. 2[18].

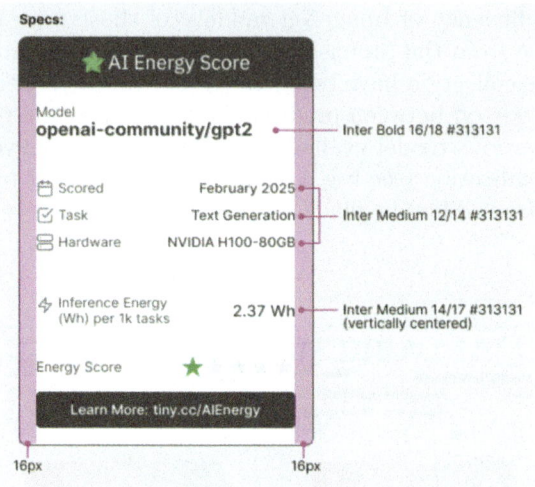

Fig. 2. AI Energy Score Model Card

Some LLMs list their energy consumption when publishing the model. *Llama 4* reduces the energy consumption and reports it for the new multimodal and multilingual versions *Scout* and *Meverick*. Model training utilized 7.38M GPU hours of computation on H100-80GB type hardware and the estimated location-based greenhouse gas emissions were 1,999 tons CO2 equivalents for training. Moreover, Meta states that it maintains net zero greenhouse gas emissions in its global operations[19].

In April 2025, I asked *Qwen* about its energy consumption and it answered with approximately 1 W hour. For the 32B *OpenHands* model it answered that the coding demands a longer context of the tokens and a higher precision. In addition, iterations and software-specific knowledge makes the task more difficult. Moreover, the inference time is probably longer. As opposed to 20–50 tokens for answering an easy question, coding could well need 100–500 tokens. The answer does not take into account the particular computing center with its water and energy consumption and the source of the energy, be it from solar, wind, or other renewable energies. This should be declared for the particular use of the chat. Otherwise, the ecological footprint of the basic regular query is not transparent.

There are now frameworks reporting energy consumption for several ML approaches and LLMs. This allows companies and users to select those with a

---

[18] https://huggingface.github.io/AIEnergyScore/ accessed the 10th of April 2025.
[19] https://huggingface.co/meta-llama/Llama-4-Scout-17B-16E-Instruct.

beneficial trade-off of quality and resource awareness. In the end, this was the intent for starting the *Care Labels* approach and the *AIEnergyScore*. However, there is a grain of salt. It has been shown that consumers are willing to pay a price for explainable but not for green AI [30]. Let us hope that developers will continue to compete for the most resource-aware algorithms!

5   Conclusion

This chapter has given an overview of, both, the IoT oriented AI and the LLMs regarding their resource consumptions. AI research has designed and implemented algorithms that are optimized for using less resources. Modern hardware and ML models in concert reduce inference time, memory needs, and, hence, the CO_2 output. Methods for measuring, testing, and reporting models have been developed and brought into a form that users can easily understand. It is now possible to choose ML models with a minimal footprint. This good news from science need to be accomplished, though. We all must take action to reduce our ecological footprint. Scientists, users, the public must push the governments to regulate AI such that we don't burn down our planet for funny pictures.

Acknowledgments. This chapter was written by me during my retirement time. It builds on the work that we did without funding when I was still heading the ls8 lab at the TU Dortmund University, computer science department. My special thanks go to Helena Kotthaus, Raphael Fischer, Danny Heinrich, Lukas Heppe, Matthias Jakobs, Sascha Mücke, Andreas Pauly, and Nico Piatkowski for the deep discussions, many experiments, and coding.

Disclosure of Interests. The author has no competing interests to declare that are relevant to the content of this article.

References

1. Ali, M., et al.: Tokenizer choice for LLM training: negligible or crucial? arXiv:2310.08754 (2024)
2. Alkhoury, F., Welke, P.: Splitting stump forests: tree ensemble compression for edge devices. In: Proceedings of the Discovery Science (2024). https://www.mlai.cs.uni-bonn.de/en/paper/alkhoury/splitting_stump_forests.pdf
3. Arnold, M., et al.: Factsheets: increasing trust in AI services through supplier's declarations of conformity. IBM J. Res. Dev. (2019). https://doi.org/10.1147/JRD.2019.2942288
4. Brown, T.B., et al.: Language models are few-shot learners. arXiv (2020). https://arxiv.org/abs/2005.14165
5. Buschjaeger, S., Morik, K.: Joint leaf-refinement and ensemble pruning through L1 regularization. Data Min. Knowl. Discov. **37** (2022). https://doi.org/10.1007/s10618-023-00921-z
6. Buschjäger, S., Morik, K.: Decision tree and random forest implementations for fast filtering of sensor data. IEEE Trans. Circuits Syst. I: Regular Pap. **65-I**(1), 209–222 (2018). https://doi.org/10.1109/TCSI.2017.2710627

7. Buschjäger, S., Pfahler, L., Buss, J., Morik, K., Rhode, W.: On-site gamma-hadron separation with deep learning on FPGAs. In: Dong, Y., Mladenić, D., Saunders, C. (eds.) ECML PKDD 2020. LNCS (LNAI), vol. 12460, pp. 478–493. Springer, Cham (2021). https://doi.org/10.1007/978-3-030-67667-4_29

8. Castano, J., Martinez-Fernandez, S., Franch, X., Bogner, J.: Exploring the carbon footprint of hugging faces ml models: a repository mining study. arXiv:2305.11164 (2023)

9. Dalleiger, S., Vreeken, J., Kamp, M.: Federated binary matrix factorization using proximal optimization. In: Proceedings of the AAAI 2025 (2025). https://arxiv.org/abs/2407.01776

10. DeepSeek: DeepSeek-r1: Incentivizing reasoning capability in LLMs via reinforcement learning. GitHub (2025). https://github.com/deepseek-ai/DeepSeek-R1/blob/main/README.md

11. Deiseroth, B., Brack, M., Schaminowski, P., Kersting, K., Weinbach, S.: T-free: tokenizer-free generative LLMs via sparse representations for memory-efficient embeddings. arXiv (2024). https://arxiv.org/pdf/2406.19223

12. Di, S., et al.: A survey on error-bounded lossy compression for scientific datasets. arXiv (2025). https://arxiv.org/abs/2404.02840

13. Faiz, A., et al.: LLMCarbon: modeling the end-to-end carbon footprint of large language models. In: 12th International Conference on Learning Representations ICLR (2024). https://openreview.net/forum?id=aIok3ZD9to

14. Fischer, R., Liebig, T., Morik, K.: Towards more sustainable and trustworthy reporting in machine learning. Data Min. Knowl. Discov. (2024)

15. Fischer, R., Wischnewski, M., van der Staay, A., Poitz, K., Janiesch, C., Liebig, T.: Bridging the communication gap: evaluating AI labeling practices for trustworthy AI development. arXiv (2025). https://arxiv.org/abs/2501.11909

16. Gala, G., de Campos, C., Peharz, R., Vergari, A., Quaeghebeur, E.: Probabilistic integral circuits. In: Proceedings of the International Conference on Artificial Intelligence and Statistics. PMLR (2024). https://proceedings.mlr.press/v238/gala24a.html

17. Georgiou, S., Kechagia, M., Louridas, P., Spinellis, D.: What are your programming language's energy-delay implications? In: Proceedings of the 15th International Conference on Mining Software Repositories. ACM (2018)

18. Gomez, A.: On-demand communication with the batteryless mirocard: demo abstract. In: Proceedings of 18th Conference on Embedded Networked Sensor Systems (2020)

19. Gomez, A., Suter, L., Mayer, S.: Zero-Power/Low-Power Sensing, pp. 47–70. De Gruyter (2022). sfb_book1_2.3. https://doi.org/10.1515/9783110785944-002

20. Guru, D., Hardt, M.: ARM cluster for performant and energy-efficient storage. In: Lässig, J., Kersting, K., Morik, K. (eds.) Computational Sustainability, pp. 265–276. Springer (2016)

21. Hähnel, M., Döbel, B., Völp, M., Härtig, H.: Measuring energy consumption for short code paths using RAPL. SIGMETRICS Perform. Eval. Rev. 40(3), 13–17 (2012).https://doi.org/10.1145/2425248.2425252

22. Henderson, P., Hu, J., Romoff, J., Brunskill, E., Jurafsky, D., Pineau, J.: Towards the systematic reporting of the energy and carbon footprints of machine learning. J. Mach. LOIearn. Res. 21(248), 1–43 (2020). http://jmlr.org/papers/v21/20-312.html

23. Heppe, L., Kamp, M., Adilova, L., Piatkowski, N., Heinrich, D., Morik, K.: Resource-constrained on-device learning by dynamic averaging. In: ECML PKDD 2020 Workshops, pp. 129–144. Springer, Cham (2020)

24. Hintemann, R., Hinterholzer, S., Progni, K.: Rechenzentren in Deutschland, aktuelle Marktentwicklungen 2024 (bitkom e.v.) (2024). https://www.bitkom.org/Bitkom/Publikationen/Studie-Rechenzentren-in-Deutschland

25. Hohman, F., et al.: TALARIA: interactively optmizing machine learning models for efficient inference. arXiv (2024)

26. Huang, W., et al.: BiLLM: pushing the limit of post-training quantization for LLMs. arXiv (2024). arXiv: 2402.04291

27. Huang, W., et al.: How good are low-bit quantized LLaMA3 models? An empirical study. arXiv (2024). https://arxiv.org/pdf/2404.14047

28. Hubara, I., Courbariaux, M., Soudry, D., El-Yaniv, R., Bengio, Y.: Binarized neural networks. In: Proceedings of the NIPS, pp. 4107–4115 (2016)

29. IPCC: Global warming of 1.5 celsius. Special Report (2023). https://www.ipcc.ch/sr15/

30. Koenig, P., Wurster, S., Siewert, M.: Consumers are willing to pay a price for explainable, but not for green AI. Evidence from a choice-based conjoint analysis. Big Data Soc. (2022). https://doi.org/10.1177/20539517211069632

31. Lacoste, A., Luccioni, A., Schmidt, V., Dandres, T.: Quantifying the carbon emissions of machine learning (2019). arXiv:1910.09700

32. Lässig, J., Kersting, K., Morik, K.: Computational Sustainability. Springer (2016). https://doi.org/10.1007/978-3-319-31858-5

33. Leeuwen, M., Siebes, A.: Streamkrimp: detection change in data streams. In: Daelemans, W., Goethals, B., Morik, K. (eds.) Proceedings of the ECML PKDD, pp. 672–687. Springer, Heidelberg (2008)

34. Li, P., Yang, J., Islam, M., Ren, S.: Making AI less thirsty: uncovering and addressing the secret water footprint of AI models. arXiv (2025). https://arxiv.org/pdf/2304.03271

35. Luccioni, A.S., Strubell, E., Crawford, K.: From efficiency gains to rebound effects: the problem of Jevons' paradox in AI's polarized environmental debate (2025). https://arxiv.org/abs/2501.16548

36. Luccioni, A.S., Viguer, S., Ligozat, A.L.: Estimating the carbon footprint of bloom, a 176b parameter language model. arXiv (2022)

37. Marwedel, P.: Embedded System Design - Embedded Systems Foundations of Cyber-Physical Systems, and the Internet of Things. Springer, 4 edn. (2021). https://doi.org/10.1007/978-3-030-60910-8

38. Masoudinejad, M., Buschhoff, M.: PhyNetLab Test Bed, pp. 34–46. De Gruyter (2022). https://doi.org/10.1515/9783110785944-002. sfb_book1_2.2

39. Meadows, D.H., Meadows, D., Randers, J., III, W.B.: The Limits of Growth: A Report for the Club of Rome's Project on the Predicament of Mankind. Macmillan (1979)

40. Mitchell, M., et al.: Model cards for model reporting. In: Proceedings of Conference on Fairness, Accountability, and Transparency, FAT* 2019, pp. 220–229. ACM (2019). https://doi.org/10.1145/3287560.3287596

41. Morik, K., Bhaduri, K., Kargupta, H.: Introduction to data mining for sustainability. Data Min. Knowl. Disc. **24**(2), 311–324 (2012)

42. Morik, K., Chen, J.J.: Introduction, pp. 1–14. De Gruyter (2022). https://doi.org/10.1515/9783110785944-001. sfb_book1_1

43. Morik, K., et al.: Yes we care! - certification for machine learning methods through the care label framework. Front. Artif. Intell. (2022). https://www.frontiersin.org/articles/10.3389/frai.2022.975029/full

44. Morik, K., Marwedel, P. (eds.): Machine Learning under Resource Constraints - Fundamentals. De Gruyter (2022). https://doi.org/10.1515/9783110785944. sfb_book1

45. Patterson, D., et al.: Carbon emissions and large neural network training. arXiv (2021). https://arxiv.org/abs/2104.10350

46. Pereira, R., et al.: Ranking programming languages by energy efficiency. Sci. Comput. Programm. (2021). https://doi.org/10.1016/j.scico.2021.102609

47. Piatkowski, N.: Integer Exponential Families, pp. 408–425. De Gruyter (2022). https://doi.org/10.1515/9783110785944-009. sfb_book1_9.1

48. Piatkowski, N., Lee, S., Morik, K.: Integer undirected graphical models for resource-constrained systems. Neurocomputing **173**(1), 9–23 (2016). http://www.sciencedirect.com/science/article/pii/S0925231215010449

49. Piatkowski, N., Morik, K.: Fast stochastic quadrature for approximate maximum-likelihood estimation. In: Proceedings of 34th Conference on UAI (2018)

50. Piatkowski, N., et al.: Generative machine learning for resource-aware 5G and IoT systems. In: IEEE International Conference on Communications ICC Workshops, pp. 1–6 (2021). https://doi.org/10.1109/ICCWorkshops50388.2021.9473625

51. Pihkola, H., Hongisto, M., Apilo, O., Lasanen, M.: Evaluating the energy consumption of mobile data transferfrom technology development to consumer behaviour and life cycle thinking. Sustainability **10**(7) (2018)

52. Ren, X., El-Kishky, A., Wang, C., Tao, F., Voss, C.R., Han, J.: ClusType: effective entity recognition and typing by relation phrase-based clustering. In: Cao, L., Zhang, C., Joachims, T., Webb, G.I., Margineantu, D.D., Williams, G. (eds.) Proceedings of 21th ACM SIGKDD, pp. 995–1004. ACM (2015). http://doi.acm.org/10.1145/2783258.2783362

53. Rolnick, D., et al.: Tackling climate change with machine learning. In: Workshop at ICML: Climate Change – How Can AI help? (2019). https://arxiv.org/pdf/1906.05433

54. Schwartz, R., Dodge, J., Smith, N.A., Etzioni, O.: Green AI. arXiv preprint arXiv:1907.10597 (2019)

55. Siebes, A., Vreeken, J., van Leeuwen, M.: Item sets that compress. In: Ghosh, J., Lambert, D., Skillicorn, D.B., Srivastava, J. (eds.) Proceedings of 6th SIAM International Conference on Data Mining. pp. 395–418 (2006)

56. Sigrist, L., Gomez, A., Lim, R., Lippuner, S., Leubin, M., Thiele, L.: Measurement and validation of energy harvesting IoT devices. In: 2017 Design, Automation & Test in Europe Conference & Exhibition (DATE), pp. 1159–1164 (2017)

57. Strubell, E., Ganesh, A., McCallum, A.: Energy and policy considerations for deep learning in NLP. In: Proceedings of the 57th Annual Meeting of the Association for Computational Linguistics (ACL) (2019). https://arxiv.org/abs/1906.02243

58. Strubell, E., Ganesh, A., McCallum, A.: Energy and policy considerations for modern deep learning research. In: The 34th AAAI, pp. 13693–13696. AAAI Press (2020). https://aaai.org/ojs/index.php/AAAI/article/view/7123

59. Tomaras, D., Buschjäger, S., Kalogeraki, V., Morik, K., Gunopulos, D.: STRATA: random forests going serverless. In: Proceedings of the International Middleware Conference. ACM (2024). https://doi.org/10.1145/3652892.3654791

60. Touvron, H., et al.: LLaMA: open and efficient foundation language models. arXiv (2023). https://arxiv.org/pdf/2302.13971

61. Vreeken, J., van Leeuwen, M., Siebes, A.: KRIMP: mining itemsets that compress. Data Min. Knowl. Discov. **23**, 169–214 (2011)

62. Wang, R., et al.: Optimizing large language model training using FP4 quantizati. arxiv (2025). https://arxiv.org/abs/2501.17116

63. Yang, A., et al.: Qwen2 technical report. arXiv:2407.10671 (2024)

Evidence Accumulation as a Combining Rule for Decision Forests

Hendrik Blockeel[1,2]([✉]), Daan Van Wesenbeeck[1,2], and Andreh Al Warrad[3]

[1] Department of Computer Science, KU Leuven, Leuven, Belgium
hendrik.blockeel@kuleuven.be
[2] Leuven.AI, Leuven, Belgium
[3] Applied Informatics, KU Leuven, Leuven, Belgium

Abstract. Much is still unknown about how to best combine probabilistic predictions in classifier ensembles. In this paper, we propose a combining rule called Evidence Accumulation (EVA) and study it in the context of ensembles of decision trees. We identify a key factor that determines the performance of EVA relative to the commonly used averaging: namely, the extent to which the combined probabilities represent epistemic, rather than aleatoric, uncertainty. This insight leads to the expectation that EVA should work better than averaging for shallow or highly randomized trees. It further leads to a view of evidence-accumulating forests as a bridge between two very different types of classifiers, namely, Naive Bayes and decision forests.

Keywords: Combining rules · probabilistic reasoning · decision trees · ensembles · Naive Bayes · information fusion

Foreword: A Personal Note to Arno Siebes

Dear Arno,

It is my firm belief that scientists should explain their work as clearly as they can. That often implies spending many hours designing the right concepts, terminology, notation, etc. so that the actual innovation can be explained with maximum clarity, precision and conciseness. This goes directly against the "impress by obfuscation" that seems to plague machine learning these days [13], and unfortunately it often feels like shooting yourself in the foot: all too often, reviewers reject work because it looks easy. They do not realize that finding a simple solution can be harder than finding a complex one: in fact, it often involves a detour via the latter [18].

For me, you embody like no one else the objective of making your work maximally understandable. Here's the quintessential Arno talk, as I remember it: You start with something so simple you could be talking to secondary school students. From there, you build up your talk, in small steps that are just as simple. And then, at slide 20 or so, I suddenly realize that I've learned something really interesting and non-trivial, and I have no idea when the shift from "this is easy" to "this is brilliant!" really happened.

© The Author(s), under exclusive license to Springer Nature Switzerland AG 2026
M. van Leeuwen and J. Vreeken (Eds.): Arno Siebes Festschrift, LNCS 16067, pp. 127–144, 2026.
https://doi.org/10.1007/978-3-032-03028-3_8

It all looks so simple when you explain it. And you make no effort to diminish that impression: quite the contrary. I remember you calling out "Eitje, toch?" during a talk (not sure why you said that in Dutch, but I remember it because to a Flemish person it sounds very Dutch indeed). Apart from all else, it shows the remarkable modesty that characterizes you. The way you present things, you're not coming up with a complex solution, you're just showing how simple the problem is.

I feel privileged to dedicate this paper to one of the most exemplary computer scientists I know. Thank you for teaching the world so many interesting things, for being a role model, for sharing so much *gezelligheid*. As you are retiring, I am confident you will find more outlets for your creativity and kindness. I wish you many happy years ahead!

Yours truly,
— Hendrik

1 Introduction

When reasoning on the basis of uncertain information, AI systems often express this uncertainty using probabilities. A natural question is then: if we have different threads of reasoning that each lead to a conclusion with a certain probability, how do we combine these probabilities? This is often done using so-called combining rules. A classic example is the noisy-or rule [14], but many other exist.

In machine learning, the concept of combining information is crucial in the context of ensemble methods. An ensemble is a model that makes predictions by combining individual predictions of multiple component models into one final prediction. When the component models are decision trees, such ensembles are often called forests. Random Forests [3] and XGBoost [5] are well-known examples of methods that learn forests. In this text, we focus on forests where each individual tree directly predicts the target variable, as in bagging or random forests (and unlike, e.g., gradient boosting, where they improve previous predictions). The forest's prediction is obtained by aggregating the individual models' predictions, for instance by computing their mean (regression) or mode (classification).

Probability estimation trees (PETs) [15] are a variant of classification trees that associate probabilities with their predictions. Ensembles of PETs combine the probabilistic predictions of the models into a single probabilistic prediction, using some kind of combining rule. Very often, the mean is used, but other combining rules have been explored, for PETs [4,6] or for probabilistic classifiers in general [2,10,17]. Many of these studies set out to improve upon the use of the mean, using a well-motivated alternative, but with limited empirical success.

In this paper, we discuss a combining rule that we call Evidence Accumulation (EVA). The rule is motivated by the desire to *accumulate* the evidence provided by multiple experts. To see why this is needed, consider this example: You got sick after going to some restaurant and suspect the food was to blame, but you are not sure; you estimate the chance at 50%. You next find out your friend ate there on the same evening and also got sick, and thinks there is a 50% chance it was the food. Putting the evidence together strongly increases the probability that it was indeed the food; but averaging your and your friend's probability estimates does not have this effect. Clearly, in this case, averaging is not the right thing to do. As we show in the next section, there is no generally correct formula for combining evidence. Additional assumptions must therefore be made. EVA assumes class-conditional independence (as in Naive Bayes).

After introducing EVA, we discuss how it can be used in the context of ensembles of PETs; we call the resulting model EAFs (Evidence Accumulating Forests). The specific structure of PETs leads to an important observation: the arguments in favor of specific combining rules rely on the interpretation of the probability estimates, and more specifically, whether the estimated probabilities are interpreted as expressing aleatoric or epistemic uncertainty. We argue that the standard combining rule (averaging) makes more sense for the aleatoric interpretation, while EVA makes sense for the epistemic interpretation. This leads to the hypothesis that the mean should work better when combining deep trees (where the epistemic uncertainty is smaller), while EVA is more suitable for combining shallow trees.

We further show that on datasets with nominal attributes, the well-known Naive Bayes classifier is equivalent to a specific type of EAF. EAFs can therefore be seen as an extension of Naive Bayes in which the conditional independence conditions are relaxed (and this relaxation is learned from data).

The paper ends with some empirical results related to the above observations, a discussion of related work, and conclusions.

2   Evidence Accumulation

2.1   Sketching the Problem

In the following, we will use set notation for probabilistic events (as is commonly done in probability theory). The knowledge of an expert is expressed as the set of possible worlds that are consistent with the expert's knowledge.

Imagine a set of experts E_i is available, each with their own knowledge. Let K_i be the set of possible worlds consistent with the knowledge of E_i. If there are multiple experts E_i, the set of worlds consistent with all their knowledge is $\bigcap_i K_i$. If expert E_i, given the knowledge to their disposal, estimates the probability of A as $P(A|K_i)$, what is then the probability of A given the combined knowledge of all experts, $P(A|\bigcap_i K_i)$? The following examples illustrate questions of this nature.

Example 1. Assume 30% of all Belgians, and 100% of all engineers, know what the derivative of $\sin(x)$ is. What is the probability that a randomly chosen Belgian engineer knows it? We have $P(K|B) = 0.3$ and $P(K|E) = 1$, with K, B, and E standing for knowing the answer, being Belgian, and being an engineer. It is intuitively clear that $P(K|B \cap E) = 1$.

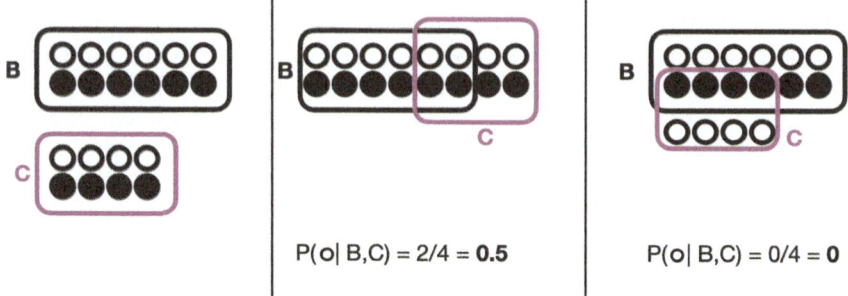

Fig. 1. B and C are sets of black and white balls. Assume A expresses the event of a ball being white. Assuming uniform distributions over B and C, $P(A|B) = P(A|C) = 0.5$. The value of $P(A|B \cap C)$ depends on how sets B and C overlap. In the middle, $B \cap C$ has 4 balls, 2 white and 2 black, so $P(A|B \cap C) = 0.5$. The right shows a situation where $B \cap C$ contains only black balls, hence $P(A|B \cap C) = 0$. If we redefine A as drawing a black ball, then $P(A|B \cap C) = 1$ in the right picture.

Example 2. A burglary was committed, and only 5 people could have done it; call them A, B, C, D, E. Now, witness F knows for sure that A and B didn't, because they were with him at the time; whereas witness G knows for sure that D and E did not do it because she was with them. Combining the evidence available to F and G, it is clear that C must have done it: this is the only possibility left. That is, $P(C|F \cap G) = 1$.

Now let us view both witnesses as experts with complementary information, and assume they express their knowledge in a probabilistic manner. Assuming a uniform prior as to who might be guilty, F will assign $P(A) = P(B) = 0$ and $P(C) = P(D) = P(E) = 1/3$. G will similarly state $P(D) = P(E) = 0$ and $P(A) = P(B) = P(C) = 1/3$. So we have $P(C|F) = 1/3$, $P(C|G) = 1/3$, and $P(C|F \cap G) = 1$. What formula relates $P(C|F \cap G)$ to $P(C|F)$ and $P(C|G)$?

These examples show that $P(A|B \cap C)$ does not in general equal the sum, product, maximum, or any simple function of $P(A|B)$ and $P(A|C)$. In fact, there exists no function that computes $P(A|B \cap C)$ from $P(A|B)$ and $P(A|C)$: the latter two do not uniquely determine the former one. Figure 1 illustrates how for $P(A|B) = 0.5$ and $P(A|C) = 0.5$, $P(A|B \cap C)$ can still vary from 0 or 1.

Thus, $P(A|\bigcap_i B_i)$ is not uniquely determined by $P(A|B_i)$, $i = 1, \ldots, k$.

2.2   Conditional Independence of Experts

Under specific assumptions, the $P(A|B_i)$ do jointly determine $P(A|\bigcap_i B_i)$. For instance, when the B_i are conditionally independent of each other, given A,

$$P(A|\bigcap_i B_i) = P(\bigcap_i B_i|A) \cdot \frac{P(A)}{P(\bigcap_i B_i)} \qquad\qquad \text{Bayes' rule}$$

$$= \Big(\prod_i P(B_i|A)\Big) \cdot \frac{P(A)}{P(\bigcap_i B_i)} \qquad\qquad \text{cond. indep.}$$

$$= \Big(\prod_i \frac{P(B_i|A)}{P(B_i)}\Big) \cdot \frac{\prod_i P(B_i)}{P(\bigcap_i B_i)} \cdot P(A) \qquad\qquad \times \frac{\prod_i P(B_i)}{\prod_i P(B_i)}$$

$$= \Big(\prod_i \frac{P(A|B_i)}{P(A)}\Big) \cdot \frac{\prod_i P(B_i)}{P(\bigcap_i B_i)} \cdot P(A) \qquad\qquad \text{Bayes' rule}$$

$$= \beta \cdot P(A) \cdot \Big(\prod_i \frac{P(A|B_i)}{P(A)}\Big) \qquad\qquad \text{renaming } \frac{\prod_i P(B_i)}{P(\bigcap_i B_i)} \text{ to } \beta$$

This result allows for an interesting interpretation. The probability of A, given the combined evidence $\bigcap_i B_i$, is obtained by starting with the unconditional $P(A)$ multiplied by a multiplicative bias β that expresses (unconditional) dependences among experts, and then, for each expert, multiplying the probability by a "lift factor" $P(A|B_i)/P(A)$, the ratio of the posterior probability (given the expert's knowledge) to the prior probability. The rule thus accumulates the evidence available to different experts, which is why we call it **evidence accumulation** (EVA).

While presented here in the more general context of events A and B_i, the rule is easily rewritten for stochastic variables and the values they take:

$$P(Y = y|\bigwedge_i X_i = x_i) = \beta \cdot P(Y = y) \cdot \prod_i \frac{P(Y = y|X_i = x_i)}{P(Y = y)} \qquad (1)$$

About the multiplicative bias β, we note the following:

- β indicates how much less likely the observed combination of X_i values is, compared to if the X_i were independent. It thus quantifies unconditional independence (as opposed to class-conditional independence, which was used to derive the formula).
- β cannot be computed directly, but can be found through normalization if needed.
- Equation 1 may seem to imply that when an additional expert is added to the expert pool, the conditional probability of A changes by that expert's lift factor. But that is inaccurate, as expanding the expert pool will typically also change β.

3   Evidence Accumulation and Forests

For classification, bagging and random forests typically use voting to combine predictions. A natural extension is to weigh the votes based on class proportions in the leaves (e.g., if a tree sorts x into a leaf with 70% positives, it would cast a positive vote with a weight of 0.7 and a negative with a weight of 0.3).

In probability estimation trees (PETs), each such proportion is interpreted as an estimated class probability, and the prediction of the ensemble as a whole is interpreted likewise. The prediction of the ensemble is then the mean of the predictions of the components. The interpretation as a class probability has led researchers to investigate probabilistic motivations for using this mean, and possible alternatives. Below, we argue that the probability estimates provided by the leaves can be interpreted in different ways, while the motivation for a specific combining rule often implicitly assumes one specific interpretation.

3.1   Interpretation of the Component Probabilities

Consider a single PET. For simplicity of exposition, we assume binary classification: given an instance, the PET predicts the probability that it is positive. Typically, this probability is estimated as the proportion of positives p in the leaf L that x is sorted into.

But there are at least two different conditional probabilities for which we can see p as an estimator. For clarity, we here adopt the convention of using a capital for a stochastic variable and the lowercase equivalent for its value. Using Y with domain $\{0, 1\}$ for the class, and X (with as domain the Cartesian product of n nominal attribute domains) for the input, p can be viewed as an estimate for

- $P(Y = 1 | X = x)$: the probability that the class of x is positive, given the attribute values of x
- $P(Y = 1 | X \in L)$: the probability that the class of x is positive, given that x got sorted into leaf L

These are two very different probabilities. The difference between them is related to the difference between aleatoric and epistemic uncertainty:

- $P(Y = 1 | X = x)$ is purely aleatoric: we assume that different occurrences of exactly the same x can have different labels; $P(Y = 1 | X = x)$ is the expected relative frequency of label 1 among these.
- $P(Y = 1 | X \in L)$, on the other hand, has an epistemic component: this probability is conditioned, not on knowing all attribute values of X, but only on knowing that $X \in L$. It is the expected relative frequency of label 1 among all instances X sorted into L.

Example 3. Assume we have a dataset with the outcome of die rolls; each die is described as 6-sided (denoted c, for cube) or 8-sided (o, octahedron), and red (r) or blue (b). So the domain of X is $\{c, o\} \times \{r, b\}$ and that of Y is $\{1, 2, 3, 4, 5, 6, 7, 8\}$. Consider a tree of depth 1 that splits on shape. In the c leaf, all x have the same probability of label 5 (namely $1/6$), and we expect p, the relative frequency of label 5 in the leaf, to be near $1/6$. This probability is purely aleatoric. If we instead split based on color, then in the b leaf, label 5 might have a relative frequency p of $7/48$, which is a good estimate of $P(Y = 5|color(X) = b)$ (converging to it as more instances get sorted into the leaf), but a less good estimate of $P(Y = 5|X = x)$ in the subpopulation of blue dice (the latter being either $6/48$ or $8/48$, depending on x).

3.2   Interpreting the Combining Rules

The difference between these interpretations of p is important when we consider what happens when combining the predictions of multiple PETs into one probabilistic prediction. If we view the ensemble as a black box, its prediction can only be interpreted as $P(Y = 1|X = x)$: that is the only interpretation that makes sense if we do not consider the internal structure of the model. Let us call this probability π.

Now, let us first interpret each prediction p_i made by an individual tree as an estimate of $\pi_i = P(Y = 1|x \in L_i)$ with L_i the leaf x is sorted into. We have $E(p_i) = \pi_i$ and $Var(p_i) = \pi_i(1 - \pi_i)/n_i$, with n_i the number of training instances in the leaf (binomial distribution). As we go down a tree, the number of cases covered by a node decreases and the number of conditions defining the node grows. A deeper leaf therefore typically has a smaller n_i, which means p_i can deviate more from π_i; but more conditions defining it, which means the condition $x \in L_i$ becomes more similar to $X = x$, and π_i gets closer to π.

Now if we see p_i as an estimate of π, then if we assume all p_i are unbiased, it makes sense to average them: their mean is then an unbiased estimator of π with smaller variance. But this condition is not fulfilled: since $E(p_i) = \pi_i$, averaging the p_i estimates the mean of the π_i, which has no practical meaning. However, the closer the π_i are to π, the more nearly the condition is fulfilled.

We can conclude from the above that deeper down the tree, where $\pi_i \simeq \pi$ (specifically, the standard deviation of p_i from π_i dominates $|\pi_i - \pi|$), averaging makes sense; while higher up in the tree, where p_i is close to π_i but the π_i are far from π, it makes more sense to try to reconstruct π from π_i using EVA. Figure 2 visualizes the two situations.

Apart from depth, the way the tree was constructed may also have an effect. Tree learners that carefully select attributes (e.g., by maximizing information gain) basically try to minimize epistemic uncertainty. Tree learners that choose tests more randomly, like random forests or extremely randomized trees [8], do not do this. Averaging therefore makes more sense for the former, EVA for the latter.

Example 4. We continue the dice example. Assume the training set reflects a perfectly uniform distribution over the input attributes. Now consider an ensemble of two trees of depth 1, each with a different predictive attribute in the root. A blue 6-sided die ends up in the "blue" leaf of the first tree and the "cube" leaf of the second. Averaging will estimate its probability of rolling 5 as $(7/48 + 8/48)/2 = 7.5/48$. EVA will estimate it as $7/48 \cdot 1 \cdot \frac{8/48}{7/48} \cdot \beta$ which after normalization ($\beta = 1$) becomes $1/6$, the correct answer. Now consider an ensemble of two trees of depth 2: the first splits first on root, then on color; the

Shallow leaf: **Deep leaf:**

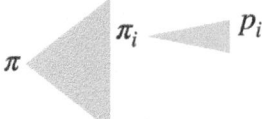   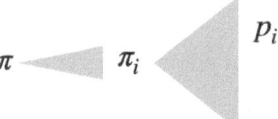

Fig. 2. A visualization of the relationship between p_i, $\pi_i = P(Y = 1|X \in L_i)$ and $\pi = P(Y = 1|X = x)$ when the i'th tree in the forest sorts instance x into leaf L_i. For shallow leaves, p_i is close to π_i and π_i can be far from π; under these assumptions, EVA, which focuses on reconstructing π from π_i, makes more sense. For deep leaves, p_i may deviate more from π_i while π_i tends to be closer to π; in this situation, averaging makes more sense.

second vice-versa. Now the blue 6-sided die ends up in two leaves with $p_i = 1/6$, so averaging predicts $1/6$; whereas EVA predicts $7/48 \cdot \left(\frac{8/48}{7/48}\right)^2 \cdot \beta$ which after normalization (we find $\beta = 1.003$) gives 0.19. For the depth-2 trees, averaging does better than EVA.

We conclude from this reasoning that averaging should work better when combining **deeper** and **less random** trees. For EVA, the opposite should hold. In the remainder of this text, we will use the acronym EAF (Evidence-Accumulating Forest) for forests that use the EVA rule.

4   EAFs and Naive Bayes

Consider a dataset with m nominal attributes. Now consider, as a special case of EAFs, an EAF with the following properties: it contains only trees of depth 1, each attribute occurs exactly once in the root of a tree, and each tree has one leaf per value of the tested attribute.

It is easily checked that this EAF is equivalent to a Naive Bayes model. Indeed, Naive Bayes estimates $P(Y = y|X = x)$ as

$$\frac{\hat{P}(Y = y) \prod_i \hat{P}(X_i = x_i|Y = y)}{\hat{P}(X = x)}$$

with $\hat{P}$ an empirical estimate of P. (The denominator is typically dropped when we only need to find the y that maximizes this.)

Using Bayes' rule on the $\hat{P}(X_i = x_i|Y = y)$ factors, this can be rewritten as

$$\frac{\hat{P}(Y = y) \prod_i \left(\hat{P}(Y = y|X_i = x_i)\hat{P}(X_i = x_i)/\hat{P}(Y = y)\right)}{\hat{P}(X = x)}$$

and with $\beta = \frac{\prod_i \hat{P}(X_i = x_i)}{\hat{P}(X=x)}$, this becomes

$$\beta \cdot \hat{P}(Y = y) \cdot \prod_i \left(\hat{P}(Y = y | X_i = x_i) / \hat{P}(Y = y) \right)$$

An EAF that contains one tree of depth 1 for each attribute X_i has exactly one leaf with $p_i = \hat{P}(Y = y | x \in L_i) = \hat{P}(Y = y | X_i = x_i)$, and thus uses exactly the same formula.

This is interesting for the following reasons:

– While there is plenty of evidence (see related work) that averaging is hard to beat in PET ensembles, at least for the case of depth-1 trees this must not be true: if it were, then Naive Bayes would work less well than averaging $\hat{P}(Y | X_i)$ over all input attributes X_i.
– It is therefore clear that which combining rule is optimal (EVA or averaging) must depend on the depth of the trees: for depth 1, EVA is best, for "typical" depths, averaging works best.
– Since Naive Bayes is equivalent to using an EAF containing m depth-1 trees, the use of deeper EAFs can be seen as an extension of Naive Bayes. While Naive Bayes is optimal under the condition that all attributes are conditionally independent given the class, EAFs are optimal under the condition that certain combinations of attribute values (namely, the ones that define the leaves of the forest) are conditionally independent given the class.

We can thus see EAFs as bridging two quite different types of models: Naive Bayes on the one hand, and decision forests on the other hand.

5   Experiments

We have run a number of experiments, mostly to check to what extent the above expectations correspond with reality.

As ensemble learner, we choose Random Forests, which is broadly considered to be state-of-the-art (XGBoost is often considered slightly better, but does not satisfy the constraint that each tree individually tries to predict the class). We use in our experiments three different versions:

– RF: the standard SciKitLearn implementation of Random Forests, which uses weighted voting
– EAF: the same except for one change: weighted voting is replaced by evidence accumulation (EVA)
– BEAF: the same as EAF, except that it constructs exactly one tree per attribute, with that attribute in its root (if additionally all trees are depth-1, this is equivalent to Naive Bayes).

Except where specified differently, 50 trees were used for each ensemble. Where results are reported for varying depths, the `max_depth` hyperparameter was used to limit the depth of the learned trees to the reported value. The plotted results are averages of ten train/test runs with 80% used for training and the remaining 20% for testing.

We use the following datasets:

- Six modified datasets from Kaggle: Wine[1] (1142 instances × 11 features); Apple[2] (3998 × 7), Student[3] (2391 × 13), Mobile[4] (1999 × 20), Bank[5] (1000 × 29) and Heart[6] (500 × 6). The datasets are modified in the sense that numerical features are made nominal (binary): values are encoded as being above/below the mean for that attribute. Note that this increases aleatoric uncertainty.[7]
- CI: synthetic datasets with 10 features that are class-conditionally independent. Each dataset is generated randomly by choosing for each instance a random class and then filling attribute values independently by drawing each randomly from a binomial distribution whose parameter depends on the class and attribute
- CD: like CI, but features are not class-conditionally independent. The parameter of each binomial distribution depends on the class, the attribute, and the value of the preceding attribute

5.1   An Implementation Detail: Laplace Smoothing

Two versions of the EVA combining rule were implemented: the version as described above, and a version that avoids extreme probabilities $(0, 1)$ by using Laplace smoothing. In a leaf with p positives and n negatives, Laplace smoothing estimates the probability of positive not as $p/(p + n)$ (the actual proportion of positives) but as $(p + 1)/(p + n + 2)$. Because Laplace smoothing never gives 0 or 1 as an estimate, it avoids the well-known problem of product-based methods (such as Naive Bayes) that a single factor of zero makes the whole product zero.

Figure 3 compares both versions in the context of EAF. The graphs clearly show that for shallow trees, both are practically equivalent, but for deeper trees, accuracy often drops when no Laplace smoothing is used. This is not unexpected: deeper leaves tend to cover fewer instances, which increases the chances

---

[1] https://www.kaggle.com/datasets/yasserh/wine-quality-dataset.
[2] https://www.kaggle.com/datasets/nelgiriyewithana/apple-quality.
[3] .../rabieelkharoua/students-performance-dataset.
[4] .../iabhishekofficial/mobile-price-classification.
[5] .../mlg-ulb/creditcardfraud.
[6] .../shantanugarg274/heart-prediction-dataset-quantum.
[7] One reviewer pointed out that with this procedure, "information from the test set leaks into the training process". This would certainly be true if the discretization were part of the learning process, but that is not the case here. Any dataset that is used for evaluation has been obtained in one way or another. In our opinion, the use of discretization, normalization, or other steps based on the whole dataset, as part of this process, does not make the resulting dataset unsuitable for evaluating learners.

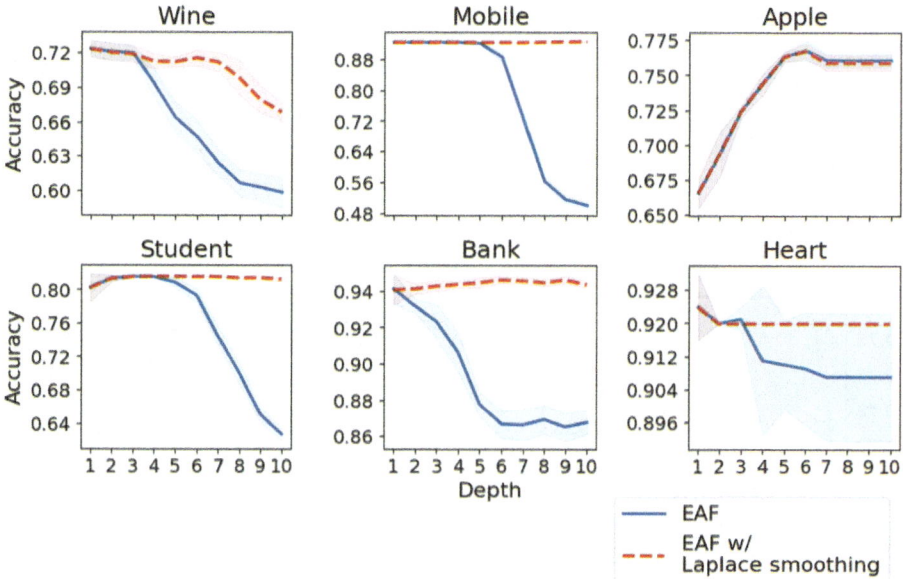

Fig. 3. Laplace smoothing prevents EVA from breaking down at deeper levels due to 0 factors. (This effect is not visible for the Apple dataset, presumably because it is large enough for leaves at depth 10 to still contain instances of both classes.

of one class being completely absent from a leaf. This in turn yields probability estimates equal to 0, and a breakdown of EVA, if no Laplace smoothing is used.

For the remainder of the experiments, we systematically use EVA with Laplace smoothing.

5.2   EVA Versus Averaging

In a first experiment, we verify our expectation that EAF outperforms RF for shallow trees, while for deeper trees the opposite holds.

Figure 4 shows that EAF indeed outperforms RF on depth 1 in most cases. In two cases (Student, Heart) the difference is very large, in two other cases rather small, and in two cases virtually nonexistent. At deeper levels, the difference mostly disappears. In two cases (Mobile and Heart), RF starts underperforming at the deepest levels; this is most likely due to overfitting (splitting up to leaves of size 1).

5.3   The Effect of Randomization

To evaluate the effect of randomness, we vary the `max_features` hyperparameter in SciKitLearn's Random Forests implementation. This hyperparameter determines the size of the random subset of attributes from which Random Forests

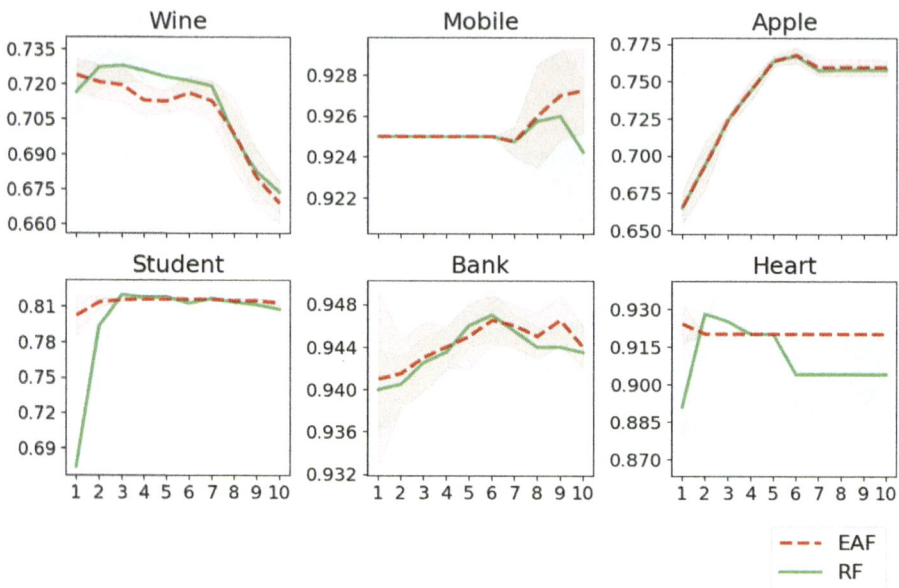

Fig. 4. Accuracy of EAF and RF, for varying maximal tree depth. EAF tends to outperform RF at level 1, but rarely for deeper levels.

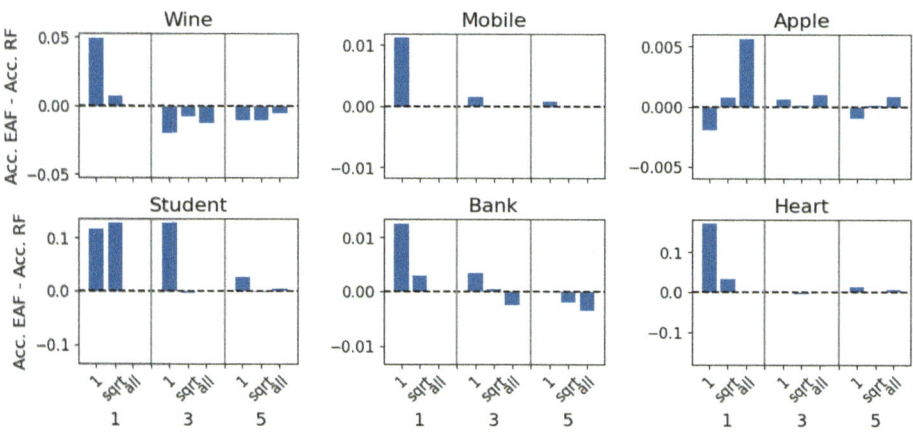

Fig. 5. The effect of randomness on the difference in performance between EAF and RF: EAF does better when there is more randomness. The effect is clearest for the datasets with the most attributes: Bank and Mobile. Apple is the only dataset where the opposite is observed, but the difference is small (note the scale of the axis).

selects the most informative one. Its default setting is `sqrt`: with d predictive attributes, the best of $\sqrt{d}$ randomly selected attributes is chosen. The smallest setting is 1 (the test is chosen entirely randomly), the largest is `all` (the best among all attributes is chosen, there is no randomness). Thus, randomness decreases in the order 1 - sqrt - all.

Figure 5 shows the difference in performance between EAF and RF. EAF generally tends to do better at depth 1, and for high randomness (the leftmost bar). The effect becomes smaller at deeper levels.

5.4   BEAF Versus RF and Naive Bayes

Figure 6 compares the predictive performance of BEAF with that of EAF and RF when these use the same number of trees as BEAF (namely, the number of attributes). In most cases, BEAF performs slightly better than the others at depth 1. The Bank dataset is a notable exception. We have not been able to find an explanation for why it behaves differently.

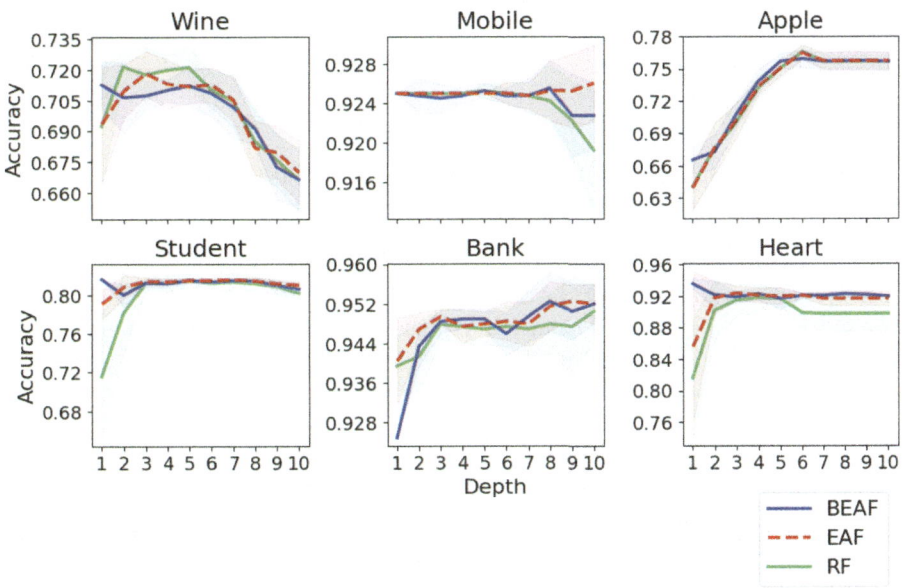

Fig. 6. Performance of BEAF compared to RF when the number of trees equals the number of attributes in the dataset.

Figure 7 compares BEAF with EAF and RF with 50 trees. Note that BEAF for depth 1 is equivalent to Naive Bayes. For four out of six datasets, better performance can be obtained by learning trees of depth greater than 1. For the Heart and Student datasets, Naive Bayes is optimal. This suggests these datasets satisfy the class-conditional independence assumption made by Naive Bayes.

We have also tested the behavior of BEAF on two synthetic datasets, one of which (CI) satisfies the Naive Bayes assumption (class-conditional independence of attributes) condition while the other (CD) does not. Figure 8 shows that for

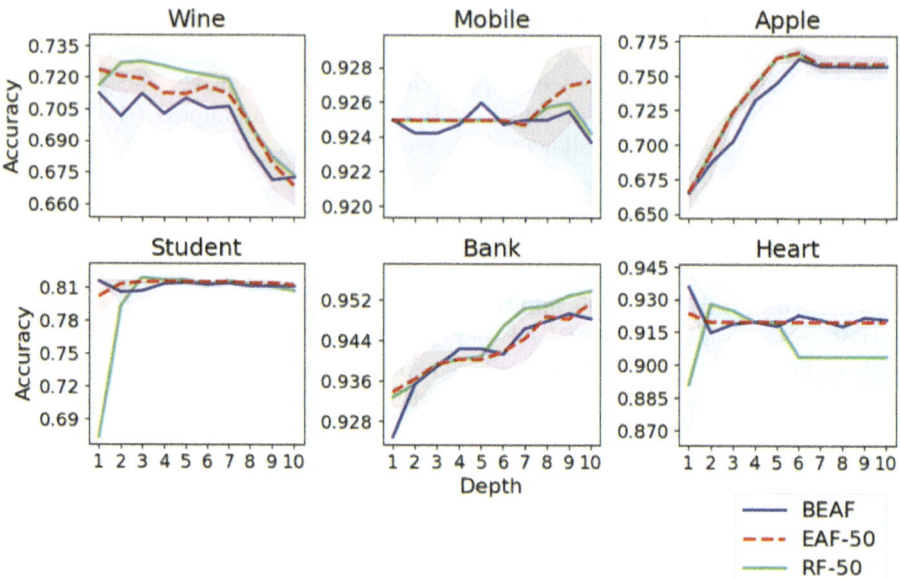

Fig. 7. Performance of BEAF compared to EAF and RF with 50 trees.

the CI dataset, best predictive performance is indeed obtained with trees of depth 1, whereas for the CD dataset, deeper trees perform better. Even when EAF and RF use more trees than BEAF, they do not outperform it for the CI dataset (as was also observed for Heart and Student).

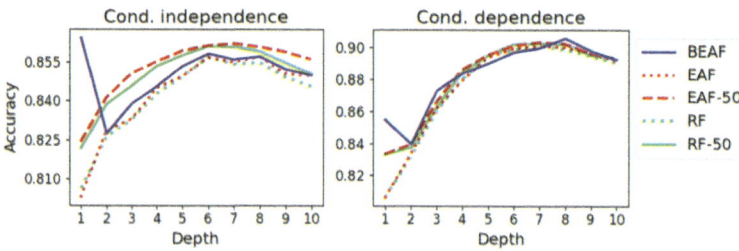

Fig. 8. Performance of BEAF compared to EAF and RF with the same number of trees, or with 50 trees (EAF-50, RF-50), on two synthetic datasets: one that satisfies class-conditional independence (CI) and one that does not (CD).

Overall, these experiments confirm our expectations that EVA has a practical advantage mostly when combining shallow trees, and that EAFs can be seen as a bridge between the simple but strongly biased Naive Bayes classifier on the one hand, and complex decision forests on the other hand. Strongly simplifying decision forests may reduce their performance to below that of Naive Bayes if the averaging rule is used; the use of EVA can avoid this.

6   Related Work

There has been a substantial amount of research on how to combine the predictions of multiple classifiers, in areas such as machine learning and information fusion. Much of this work is highly empirical, and much of it focuses on classifier ensembles in general rather than the specific case of decision forests. Excellent overviews of these topics include those by Kuncheva [12] and Kotsiantis [11].

Multiple authors have pointed out important differences between summation-based (of which averaging is a special case) and product-based combination rules for probability estimates. For instance, Hinton [9] explains how products can sharpen probability distributions, while sums cannot. A series of papers by Duin's group [10,17] is of particular interest here. Kittler et al. [10] theoretically derive a product rule and a sum rule for combining classifiers. Their product rule is essentially the same as evidence accumulation (though not interpreted in this way), while the sum rule is equivalent to averaging. They argue that the product rule should work well when the combined probability estimates are class-conditionally independent, and the sum rule should work well when *additionally* the posterior class probabilities do not deviate much from the prior. They find empirically that the sum rule works best in practice (despite relying on stricter conditions), and relate this to higher sensitivity of the product rule to differences between estimated and real probabilities. In follow-up work, Tax et al. [17] relate the conditions under which the product rule is successful to having different, uncorrelated feature spaces. In contrast, we relate the performance of EVA to the distinction between epistemic and aleatoric uncertainty. Our interpretation is compatible with Tax et al.'s: when two feature spaces are class-conditionally independent, the features of space 1 do not hold information about the features of space 2, and hence, predictions using only the former must contain epistemic uncertainty. Tax et al. find that multiplication tends to work better only in multi-class problems with "good estimates" of class probabilities (meaning mostly, no extreme estimates). The condition of "good estimates" somewhat contradicts our own finding that EVA is mostly useful when there is large epistemic uncertainty (which leads to poor estimates).

De Bock and Van den Poel [1] compare several variants of weighted averaging in the context of churn prediction. The ratio by which the probability of churn is increased (called "lift") is an important criterion in this application, and some of the weights they consider are based on this ratio. Their use of lift is motivated by the application, and they use it as weight when averaging, whereas we motivate the use of the same ratio (posterior over prior probability) by the principle of evidence accumulation, and use it in a multiplicative context.

Boström et al. [2] study "evidential" combining rules, some of which are based on Dempster-Shafer theory. Dempster-Shafer theory proposes a more sophisticated view on how to combine evidence, and does not translate in an unambiguous manner to the context of ensembles. Boström et al. do not find any method to systematically improve upon weighted voting. The connection between ensemble learning and Dempster-Shafer theory is also explored by Xu et al. [19], who integrate evidential reasoning into the ensemble learning process. They include an extensive empirical evaluation of multiple methods in this framework.

The above work does not focus on decision trees. Our focus on decision trees, and how they estimate probabilities from proportions in leaves, was instrumental in identifying the distinction between epistemic and aleatoric uncertainty as a key factor. We believe that at least part of the reason why earlier work did not lead to clear conclusions, even in the context of forests [6], is that it did not make this distinction. Our expectation that EVA should work better for shallow trees was based on this key factor, and confirmed experimentally.

Busch et al. [4] studied ensembles of random decision trees, and included EVA as a combining rule (without studying it in detail). They found that among many combining rules tried, EVA often performed best. The fact that strongly randomized decision trees were used is crucial here: standard random forests, where trees are much less random, outperformed all the approaches based on random trees. Busch et al.'s results are consistent with our analysis, in the sense that highly randomized trees contain leaves whose class entropy has not been minimized, and whose predictions therefore contain more epistemic uncertainty.

Qin [16] proposed methods that combine Naive Bayes with one or more PETs, and observed that these hybrid methods work best with shallow trees. They did not provide an explanation for this, but it is clearly in line with our analysis.

Viewing EAFs as an extension of Naive Bayes begs the question of how they compare to other extensions, such as Tree Augmented Naive Bayes (TAN) [7]. (The "tree" in TAN has nothing to do with a decision tree, but with the shape of Bayesian network that TAN is equivalent to.) This remains open for investigation.

7   Conclusions

This work was motivated by the observation that when combining probabilistic predictions by experts, the concept of accumulating evidence is relevant: when multiple pieces of evidence point towards the same conclusion, they should strengthen that conclusion. Averaging probabilities is not consistent with that intuition. We have proposed a combining rule called Evidence Accumulation that is consistent with it. Considering the individual trees in an ensemble as experts, the question is then why averaging works so well in practice, and under what conditions EVA can improve upon it. We have argued that averaging makes more sense when the uncertainty associated with the individual predictions is mostly aleatoric, and EVA makes more sense when it is mostly epistemic. This leads to the expectation that EVA should work better for shallow and more random trees. This expectation is confirmed by our experiments but also sheds new light on earlier empirical findings [4, 16].

This result has limited practical importance, in the sense that forests, and especially PET ensembles, typically use deep trees, where EVA is unlikely to be advantageous. However, there is increasing interest in using simpler models, e.g., for battery-powered devices with limited memory; in this context, our result may have practical relevance.

The main contribution of this work is that it provides novel insights. It likely explains to some extent why so many empirical comparisons of combining rules gave equivocal results: without distinguishing epistemic and aleatoric uncertainty, it is to be expected that no clear pattern can be discerned. Furthermore, it makes an interesting connection between decision forests and Naive Bayes, and suggests that shallow BEAFs can be seen as a bridge between the two that avoids their individual weaknesses.

These results offer multiple perspectives for follow-up work. First, our experiments are quite preliminary; the role of epistemic and aleatoric uncertainty could be studied in much detail, with experiments on a larger scale. Apart from that, several potential research directions seem interesting: (1) A comparison with other methods that extend Naive Bayes, such as TAN, would be a natural next step. (2) One can think of a version of decision forests that adapts its combining rule to the depth of the trees, or a hybrid combining rule that combines leaves at different depth in different ways. (3) The principle of evidence accumulation was worked out here for class-conditional independence, but there may be other sets of assumptions for which it can be implemented. (4) While the distinction between aleatoric and epistemic uncertainty is easy to make for decision trees, this is not so for classifiers in general. One wonders to what extent an analysis such as the one made here could be conducted on other types of models, and to what extent it could turn out relevant for the broader field of classifier fusion.

Acknowledgements. This research was supported by the Flemish government under the "Onderzoeksprogramma Artificiële Intelligentie (AI) Vlaanderen" programme, and by the KU Leuven Research Fund (C2E/23/007).

Disclosure of Interests. The authors have no competing interests to declare that are relevant to the content of this article.

References

1. De Bock, K.W., Van den Poel, D.: Ensembles of probability estimation trees for customer churn prediction. In: García-Pedrajas, N., Herrera, F., Fyfe, C., Benítez, J.M., Ali, M. (eds.) IEA/AIE 2010. LNCS (LNAI), vol. 6097, pp. 57–66. Springer, Heidelberg (2010). https://doi.org/10.1007/978-3-642-13025-0_7
2. Boström, H., Johansson, R., Karlsson, A.: On evidential combination rules for ensemble classifiers. In: 11th International Conference on Information Fusion, FUSION 2008, Cologne, Germany, 30 June–3 July 2008, pp. 1–8. IEEE (2008)
3. Breiman, L.: Random forests. Mach. Learn. **45**(1), 5–32 (2001)
4. Busch, F., Kulessa, M., Loza Mencía, E., Blockeel, H.: Combining predictions under uncertainty: the case of random decision trees. In: Soares, C., Torgo, L. (eds.) DS 2021. LNCS (LNAI), vol. 12986, pp. 78–93. Springer, Cham (2021). https://doi.org/10.1007/978-3-030-88942-5_7

5. Chen, T., Guestrin, C.: XGBoost: a scalable tree boosting system. In: Proceedings of the 22nd ACM SIGKDD International Conference on Knowledge Discovery and Data Mining, KDD 2016, pp. 785–794. ACM (2016)
6. Fierens, D.: On the use of combining rules in relational probability trees. In: Proceedings of the SIAM International Conference on Data Mining, SDM 2010, Columbus, Ohio, USA, 29 April–1 May 2010, pp. 397–408. SIAM (2010)
7. Friedman, N., Geiger, D., Goldszmidt, M.: Bayesian network classifiers. Mach. Learn. **29**(2–3), 131–163 (1997)
8. Geurts, P., Ernst, D., Wehenkel, L.: Extremely randomized trees. Mach. Learn. **63**(1), 3–42 (2006)
9. Hinton, G.E.: Training products of experts by minimizing contrastive divergence. Neural Comput. **14**(8), 1771–1800 (2002)
10. Kittler, J., Hatef, M., Duin, R.P.W., Matas, J.: On combining classifiers. IEEE Trans. Pattern Anal. Mach. Intell. **20**(3), 226–239 (1998)
11. Kotsiantis, S.B.: Bagging and boosting variants for handling classifications problems: a survey. Knowl. Eng. Rev. **29**(1), 78–100 (2014)
12. Kuncheva, L.: Combining Pattern Classifiers: Methods and Algorithms, 2nd edn. Wiley, Hoboken (2014)
13. Lipton, Z.C., Steinhardt, J.: Research for practice: troubling trends in machine-learning scholarship. Commun. ACM **62**(6), 45–53 (2019)
14. Pearl, J.: Probabilistic Reasoning in Intelligent Systems: Networks of Plausible Inference. Morgan Kaufmann Publishers Inc., San Francisco (1988)
15. Provost, F., Domingos, P.: Tree induction for probability-based ranking. Mach. Learn. **52**, 199–215 (2003)
16. Qin, Z.: Naive Bayes classification given probability estimation trees. In: 2006 5th International Conference on Machine Learning and Applications (ICMLA 2006), pp. 34–42 (2006)
17. Tax, D.M.J., van Breukelen, M., Duin, R.P.W., Kittler, J.: Combining multiple classifiers by averaging or by multiplying? Pattern Recogn. **33**(9), 1475–1485 (2000)
18. Weinberger, K.: The importance of deconstruction. https://slideslive.com/38938218/the-importance-of-deconstruction
19. Cong, X., Zhang, Y.Y., Zhang, W., HongQuan, Z., Zhang, Y.Z., He, W.: An ensemble learning method based on an evidential reasoning rule considering combination weighting. Comput. Intell. Neurosci. **1–17**(03), 2022 (2022)

Neighborhood-Based Collaborative Filtering Bandits

Noah Daniëls[1]([⊠])(iD) and Bart Goethals[1,2](iD)

[1] University of Antwerp, Antwerp, Belgium
{noah.daniels,bart.goethals}@uantwerpen.be
[2] Froomle, Antwerp, Belgium

Abstract. We investigate a novel collaborative filtering bandit algorithm using neighborhood-based aggregation. Our method determines pairwise user similarities from past interactions and uses them to personalize the predictions of an adapted (contextual) bandit algorithm. Personalization is achieved by reweighting past observations, thereby skewing the aggregated information towards interactions with similar users. Only taking into account similar users is not enough however and we explicitly incorporate global information to enable good performance in cold-start situations. Experiments on three real-world datasets and across different scenarios show that our algorithm performs competitive against state-of-the-art methods while being conceptually more intuitive.

Keywords: Multi-armed bandits · Recommendation

1 Introduction

The multi-armed bandit framework has garnered significant attention within the recommender systems research community in its attempts to facilitate personalized user experiences, handling changing choice sets and adapting to evolving user preferences [20]. Bandit algorithms offer a principled approach to balancing exploration and exploitation in recommendation tasks, enabling systems to learn from user feedback in real-time. Importantly, they help address the "cold-start" problem, efficiently making recommendations even when data is scarce.

With the success of collaborative filtering [13], many recent works have attempted to apply collaborative filtering techniques to bandit learning in order to improve their efficiency. For example, early works have proposed exploiting known relations between users (e.g. from a social network) to constrain bandit learning [5,27]. Other research efforts have applied online clustering techniques to learn user relations online [9,10,16–18,28], or have attempted to learn latent user and item features through the use of matrix factorization [11,12,24,25]. On the other hand, simple neighborhood approaches [19] have mostly been overlooked, despite their persistent success in collaborative filtering [2,8,22].

In this work, we present a simple neighborhood-based bandit algorithm which performs surprisingly well in a wide range of recommendation scenarios.

M. van Leeuwen and J. Vreeken (Eds.): Arno Siebes Festschrift, LNCS 16067, pp. 145–160, 2026.
https://doi.org/10.1007/978-3-032-03028-3_9

2   Background

2.1   Problem Description

In the classical multi-armed bandit problem [21], a learner is faced with a row of slot machines (arms), each with an unknown reward distribution. The goal is to maximize the total reward obtained over a series of pulls. This requires balancing the exploration of arms to learn their reward distributions with the exploitation of arms that seem to provide the highest rewards.

In recommender systems, arms represent items and pulling an arm corresponds to recommending the item to a user, with rewards indicating click feedback. In the standard setting, all users are treated equally, which means that the recommendations are not personalized. To allow for personalization, additional information about the user and/or items can be given as context. In such a setting, the learner can obtain better performance if it is able to infer the associations between contexts and rewards.

In this work, we consider two distinct contextual *scenarios*. In the "*semi-contextual*" scenario, the only contextual information is the identifier of the current user, which allows for all interactions of the same user to be linked together. In the broader "*contextual*" scenario the learner additionally has access to feature vectors describing the available items. This allows the learner to generalize to new items as in content-based recommendations.

We consider these scenarios separately because in practice it is often difficult to obtain useful item features, especially in domains where data is scarce and bandit learning is most useful (e.g. in news or online media). As the quality and availability of contextual information directly impacts performance, we investigate how various approaches handle these scenarios differently.

2.2   Notation

Formally, the problem we consider is described as follows: In each round t the learner is presented with a user identifier $u_t \in \mathcal{U}$ and a subset of available items $\mathcal{A}_t \subseteq \mathcal{A}$. In the full contextual scenario, the learner also observes $|\mathcal{A}_t|$ d-dimensional context vectors $\mathbf{x}_{t,1}, \ldots, \mathbf{x}_{t,k_t} \in \mathbb{R}^d$ associated with the available items. The learner then chooses an item $a_t \in \mathcal{A}_t$ to recommend and observes a reward $r_t \in \{0,1\}$ indicating whether the recommendation was clicked. The learner incorporates this interaction in it's model and starts the next round.

2.3   Solutions

Non-contextual. Common solutions for the classic non-contextual multi-armed bandit problem include ϵ-greedy, Upper Confidence Bound (UCB), and Thompson Sampling (TS). ϵ-Greedy is a simple approach that uses a combination of random exploration and greedy arm selection. UCB considers the upper confidence bound of am arm's expected reward, which encourages exploration of arms with uncertain rewards while exploiting arms that appear to have higher expected rewards. Thompson Sampling employs a Bayesian approach by sampling from the posterior distribution of arm rewards, selecting the arm with the highest sampled value. This results in pulling arms in proportion with the probability that they are optimal given the algorithm's current belief about the arms' distributions. In this work we focus on UCB-like algorithms, but some of our results are also applicable to TS.

Semi-contextual. For the semi-contextual scenario, one approach could be to use a separate instance of UCB per user. This allows for personalization, but it also severely limits the available information as the recommendation can depend only on previous interactions with the current user.

For this reason, various works have tried to leverage collaborative effects to improve the sample efficiency of this personalized approach. For example, if two users are expected to behave similarly, the interactions of either can be used to inform decisions for the other. This principle allows the separate instances of UCB to share feedback and learn faster. To this end, some works have focused on exploiting users relationships known a priori to the learner [5,27] while other studies have focused on clustering techniques to learn the user relationships from the interactions in an online manner [9,10,16–18]. Our work falls into the second research direction but differs in that we do not use clustering but a very simple similarity-based weighting approach.

Contextual. For the contextual scenario, the most well-known algorithm is LinUCB [7,14]. It estimates the expected reward of each arm as a linear function of the context. Thus, during round t, it estimates the regression coefficient $\mathbf{w}_t \in \mathbb{R}^d$ to compute the expected reward of arm a as $\mathbf{w}_t^\top \mathbf{x}_{t,a}$. The power of LinUCB comes from a closed form expression for the confidence bound while allowing fast learning of the regression coefficients.

As with UCB, the linear version can be applied either globally (all users treated the same) or per user. In this scenario the later option is actually viable as the contexts allow the bandit to learn from fewer interactions. Still, collaborative effects can be leveraged to further improve this baseline. In fact, all works mentioned for the semi-contextual scenario actually build on top of LinUCB instead of UCB with this scenario in mind.

3   Neighborhood Bandit

In this section, we present our neighborhood-based bandit. For ease of exposition and to motivate the algorithm, we first cover the semi-contextual version (see Algorithm 1) where only the current user is provided as context. We then consider the contextual version (see Algorithm 2) where each item is associated with a feature vector to enable faster learning.

When considering whether to recommend an item, different perspectives arise. A greedy approach recommends the item with the highest observed click-through rate (CTR). A bandit approach might also explore items with uncertain

CTR estimates. From a collaborative filtering standpoint, it is crucial to account for how similar the clicking users are to the current user. For example, if an item has a global CTR of 15% but a 25% CTR among similar users, then the personalized estimate should reflect the latter.

Our method combines these ideas: it adjusts past observations by user similarity and applies an adapted UCB algorithm on the weighted interactions, balancing exploration with personalized exploitation.

3.1   Semi-contextual Version

Algorithm 1. NUCB

Require: exploration parameter $\alpha \in \mathbb{R}_{\geq 0}$
Require: user similarity measure SIM $: \mathcal{U} \times \mathcal{U} \to [0, 1]$
Require: user similarity importance weight $\beta \in \mathbb{R}_{\geq 0}$
1: $n_{u,a} \leftarrow 0$ for all $u \in \mathcal{U}, a \in \mathcal{A}$ {impression counts}
2: $c_{u,a} \leftarrow 0$ for all $u \in \mathcal{U}, a \in \mathcal{A}$ {click counts}
3: **for** $t = 1, \ldots, T$ **do**
4:    receive user u_t to serve
5:    get set of available arms $\mathcal{A}_t \subseteq \mathcal{A}$
6:    **for** $a \in \mathcal{A}_t$ **do**
7:       compute aggregate counts $\hat{c}_{t,a}$ and $\hat{n}_{t,a}$ using (1) and (2)
8:    **end for**
9:    compute normalization factor z_t using (3)
10:   **for** $a \in \mathcal{A}_t$ **do**
11:      $p_{t,a} \leftarrow \dfrac{z_t \hat{c}_{t,a}}{z_t \hat{n}_{t,a}} + \alpha \sqrt{\dfrac{\log t}{z_t \hat{n}_{t,a}}}$ {calculate item scores}
12:   **end for**
13:   select arm $a_t = \arg\max_{a \in \mathcal{A}_t} p_{t,a}$
14:   observe payoff r_t for arm a_t
15:   $n_{u_t,a_t} \leftarrow n_{u_t,a_t} + 1$
16:   $c_{u_t,a_t} \leftarrow c_{u_t,a_t} + r_t$
17: **end for**

Our algorithm (dubbed NUCB for Neighborhood UCB) is based on running a separate instance of UCB per user. As such, NUCB maintains a count of impressions (number of recommendation) and clicks for each user-item pair. These counts are initialized to zero and increase each time a user clicks or views the recommendation of a given item (see lines 1–2 and 15–16).

When a user u_t arrives during round t, NUCB first computes aggregate click counts $\hat{c}_{t,a}$ and aggregate impression counts $\hat{n}_{t,a}$ for each available item $a \in \mathcal{A}_t$. This aggregation is based on the total count of the item plus a weighted sum over the counts of all users, so that more similar users contribute more to the total:

$$\hat{c}_{t,a} = \sum_{u' \in \mathcal{U}} (1 + \beta \text{SIM}(u_t, u')) c_{u',a} \qquad (1)$$

$$\hat{n}_{t,a} = \sum_{u' \in \mathcal{U}} (1 + \beta \text{SIM}(u_t, u')) n_{u',a} \tag{2}$$

where $\text{SIM}_U(u_t, u) \in [0, 1]$ indicates the similarity between users u_t and u, and β_U is a hyper-parameter controlling the influence of user-based similarity on the total.

We include global information by adding 1 to the similarity to combat data scarcity. When most similarities are zero (as is the case for new users) or when the counts of similar users are low (as is the case for new items), the aggregated count will mostly consist of global information. This will make the algorithm act as if it were using a single instance of UCB. As more information becomes available and the similar users start to have a significant number of interactions, the global information will no longer dominate and the algorithm will exhibit more personalization. Our experiments in Sect. 4 show the importance of this behavior.

Line 9 calculates a normalization factor to ensure that UCB does not underestimate the confidence bound. Without normalization, the aggregate counts would make it seem as if the items were impressed more often than they actually are, leading to an over confident bound and reduced exploration. Instead, the total number impression should remain unaffected and only the way they are distributed across items should change. This is achieved by multiplying the click and impression counts by the factor

$$z_t = \frac{\sum_{u \in \mathcal{U}} \sum_{a \in \mathcal{A}_t} n_{u,a}}{\sum_{a \in \mathcal{A}_t} \hat{n}_{t,a}} \tag{3}$$

Combined with aggregation, this normalization can be thought of as "skewing" the dataset of past interactions towards a dataset which contains the same number of observations, but where a greater proportion are with users that are similar to the current user.

The rest of the algorithm follows the typical formulation of UCB. Line 11 computes the value of each item as the mean reward plus an estimate of the confidence bound (which depends on the number of impressions and elapsed rounds). Line 11 also applies the normalization, which makes it clear that the CTR is unaffected and normalization is only for the benefit of the bandit algorithm and its estimate of uncertainty. Finally, line 13 selects the item with the highest value and recommends this to the user.

Because only the counts are affected by this method, our approach is not limited to UCB and can be easily applied to Thompson Sampling or other noncontextual bandit algorithm which base their decisions on impression and click counts.

Extension to Item-Based Similarity. Collaborative effects also apply to items. Given a similarity measure between items $\text{SIM}_A : \mathcal{A} \times \mathcal{A} \rightarrow [0, 1]$ and an importance weight $\beta_A \in \mathbb{R}_{\geq 0}$, the following term can be added to the aggregate click and impression counts (replacing $c_{u_t,a'}$ with $n_{u_t,a'}$ accordingly):

$$\beta_A \sum_{a' \in \mathcal{A}} \text{SIM}_A(a, a') c_{u_t, a'} \qquad (4)$$

This will add counts from the current user with similar items to the total. If a user has interacted with many similar items this can be very informative. The effect of this variant is investigated in Sect. 4.

Similarity Measures. For brevity, the presented algorithm assumes the similarity measures are simply given. In practice, they can be batch-learned online based on the interactions. An effective way of doing so is by using pairwise cosine similarity based on the click matrix $\mathbf{C} \in \mathbb{R}^{|\mathcal{U}| \times |\mathcal{A}|}$ where each entry $c_{u,a}$ is simply the click count for the given user and item. Using this approach, both the pairwise user and pairwise item similarities can be determined online.

To limit the number of similarities that need to be stored and decrease the computational cost, the aggregated counts can be computed using only the top-k most similar users and items, similar to k-nearest neighbor approaches.

3.2 Contextual Version

Algorithm 2. NLinUCB

Require: exploration parameter $\alpha \in \mathbb{R}_{\geq 0}$
Require: user similarity measure $\text{SIM} : \mathcal{U} \times \mathcal{U} \to [0, 1]$
Require: user similarity weight $\beta \in \mathbb{R}_{\geq 0}$
Require: number of features $d \in \mathbb{N}$
1: $\mathbf{M}_u \leftarrow \mathbf{I}_d \in \mathbb{R}^{d \times d}$ for all $u \in \mathcal{U}$
2: $\mathbf{b}_u \leftarrow \mathbf{0} \in \mathbb{R}^d$ for all $u \in \mathcal{U}$
3: **for** $t = 1, \ldots, T$ **do**
4:     receive user $u_t \in \mathcal{U}$ to serve
5:     get contexts $\mathbf{x}_1, \ldots, \mathbf{x}_{k_t} \in \mathbb{R}^d$ describing the available arms
6:     $\hat{\mathbf{M}}_t \leftarrow \mathbf{I}_d + \sum_{u \in \mathcal{U}} (1 + \beta \text{SIM}(u_t, u))(\mathbf{M}_u - \mathbf{I}_d)$
7:     $\hat{\mathbf{b}}_t \leftarrow \sum_{u \in \mathcal{U}} (1 + \beta \text{SIM}(u_t, u)) \mathbf{b}_u$
8:     $\hat{\mathbf{w}}_t \leftarrow \hat{\mathbf{M}}_t^{-1} \hat{\mathbf{b}}_t$
9:     $z_t \leftarrow \dfrac{||\mathbf{I}_d + \sum_{u \in \mathcal{U}} (\mathbf{M}_u - \mathbf{I}_d)||}{||\hat{\mathbf{M}}_t||}$
10:    **for** $a = 1, \ldots, k_t$ **do**
11:        $p_{t,a} \leftarrow \hat{\mathbf{w}}_t^\top \mathbf{x}_a + \alpha \sqrt{\mathbf{x}_a^\top (z_t \hat{\mathbf{M}}_t)^{-1} \mathbf{x}_a}$
12:    **end for**
13:    select arm $a_t = \arg\max_a p_{t,a}$
14:    observe payoff r_t for arm a_t
15:    $\mathbf{M}_{u_t} \leftarrow \mathbf{M}_{u_t} + \mathbf{x}_{a_t} \mathbf{x}_{a_t}^\top$
16:    $\mathbf{b}_{u_t} \leftarrow \mathbf{b}_{u_t} + r_t \mathbf{x}_{a_t}$
17: **end for**

The contextual version of our algorithm operates in much the same way as the semi-contextual version, except LinUCB is adapted instead of UCB. This means that there are no simple counters, but instead matrices $\mathbf{M}_u$ and vectors $\mathbf{b}_u$ which are incrementally updated and used to estimate the regression coefficients and confidence bound [7]. Similar to before, these matrices and vectors are aggregated based on global information and similarity with other users (see lines 6 and 7). This aggregation is similar to the clustering of bandit algorithms, except they don't use weighted sums and also don't include global information [9,10,16–18].

As with the semi-contextual version, the user similarity can be learned online based on cosine similarity. In this case, the underlying values used for this could be the $\mathbf{b}_u$ vectors as they serve a similar function to the click counts in the semi-contextual case. Compared to the semi-contextual case, the item-based extension isn't really needed here since the contextual nature already provides a way in which interactions with one item inform decision for other items.

4   Experiments

4.1   Evaluation Methodology

We use the replay evaluation method [15], which relies on a dataset of bandit feedback (i.e. containing both clicks and non-clicks) collected by recommending random items. During evaluation, each dataset entry is either discarded or kept based on whether the actions of the algorithm being evaluated matches with what is recorded. This approach provides an unbiased counterfactual view of how an algorithm would have performed if it had been used instead of the random recommendations. The downsides of this evaluation procedure are the strict requirements on the dataset for this to work well. The two public datasets that are most often used are two from Yahoo!, but unfortunately these datasets do not contain reliable user identifiers and have very few items.

Because of this limitation, we follow related work and also consider simulations based on traditional recommendation system datasets [9,10,18]. These datasets contain interactions between users and items. Under the assumption that recorded interactions represent relevant items, reward can be determined for arbitrary user-item pairs simply by looking at the dataset. The reward is 1 if the user interacted with the item and 0 otherwise.

Concretely, we operate on a fixed number of rounds. The number of rounds is 200,000 for the semi-contextual scenario and 50,000 for the contextual scenario to match related works [9,10] and because faster learning is possible through the contexts. During each round, a user is chosen uniformly at random to be the current user. The learner is presented with this user and a pool of available items. The pool consists of 1 relevant and 24 non-relevant items that are chosen at random [5]. If no relevant items remain for the current user, another user is chosen instead. After the learner selects an item, the reward is determined from the dataset and the next round begins. To make the simulation more realistic, we ensure that the item pool never contains items that were previously recommend to the current user.

Table 1. Dataset statistics after pre-processing. Showing number of users and items, context size (where applicable) and size of dataset.

Dataset	$|\mathcal{U}|$	$|\mathcal{A}|$	d	records
Yahoo! news	18,363	323	N/A	2,829,308
LastFM	1,892	17,632	25	92,834
Delicious	1,867	69,223	25	104,799

4.2 Datasets

Statistics for the used datasets can be found in Table 1.

For our evaluations using the replay method, we use the Yahoo! "Today Module" news recommendation dataset[1]. We adopted the same pre-processing used by related clustering of bandit papers and used the larger of the two versions they consider [10,17]. This pre-processing includes deriving user identities from the user features present in the dataset. We should note that this results in unrealistic "users" with thousands of interactions per day. Since this dataset contains no item features, this falls under the semi-contextual scenario.

For our simulations using traditional datasets we used LastFM and Delicious[2] [4]. We performed the same tag analysis as all clustering of bandit papers to create feature vectors [5] for the items, which allows us to use these dataset for both contextual scenarios. For the fully contextual scenario, we did not remove any users or items to align with previous experimental workflows. For the semi-contextual scenario, we kept the 1000 most occurring items because otherwise none of the algorithms were able to learn much.

LastFM and Delicious represent opposites in the way rewards depend on users. On LastFM, there are many popular artists liked by everyone, resulting in rewards that depend less on the user and more on the item. Delicious is the opposite with no popular items and each item instead being liked by only a handful of users. This makes personalization more important for Delicious.

4.3 Algorithms

We compare our algorithm (NUCB and NLinUCB) against the following baselines and state-of-the-art techniques:

- RANDOM: always selects a random item from the pool.
- UCB: the classical UCB algorithm.
- CLUB: an algorithm based on LinUCB which learns user clusters that share the same preferences. [10]

---

[1] Yahoo! Webscope dataset ydata-frontpage-todaymodule-clicks-v2_0.

[2] Both LastFM and Delicious were made available by the HetRec 2011 workshop. They were extracted from the music streaming service Last.FM and social bookmarking website Delicious.

– DYNUCB: another clustering algorithm, but which uses k-means clustering and thus requires the number of clusters as a hyperparameter. [18]
– SCLUB: a recent extension of CLUB which allows for non-uniform distribution over users. [16]
– NBANDIT: a recent multi-armed nearest-neighbor bandit algorithm which considers users as arms[3] [19].

For the semi-contextual case, the item features used by the contextual algorithms are simply one-hot encodings of the items. For the contextual scenario we replace UCB with ONE-LINUCB and MUL-LINUCB, indicating the global and individual versions of LinUCB respectively.

NBANDIT was not used on the Yahoo! dataset because there are no reliable user identifiers. Furthermore, as it cannot use item features, it was not evaluated in the contextual scenario.

For NUCB and NLINUCB we use online similarities based on the click matrix C and the $\mathbf{b}_u$ vectors as discussed previously. The similarity weights β and β_A are both set to 100. The effect of these choices is studied in Sects. 4.5 and 4.6.

For all experiments comparing the state of the art, the hyper-parameters were tuned using grid search on 20% of the total number of round T, after which the optimal parameters were frozen. This is similar to the parameter tuning procedure adopted by related works [9,10,17].

4.4   Comparison with the State of the Art

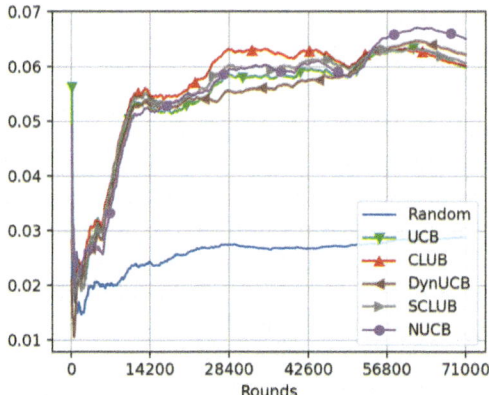

Fig. 1. Cumulative CTR of various algorithms on the Yahoo! dataset using the replay evaluation method.

---

[3] We adapted the algorithm to only consider users which have clicked at least one item in the pool, otherwise the recommendations were random.

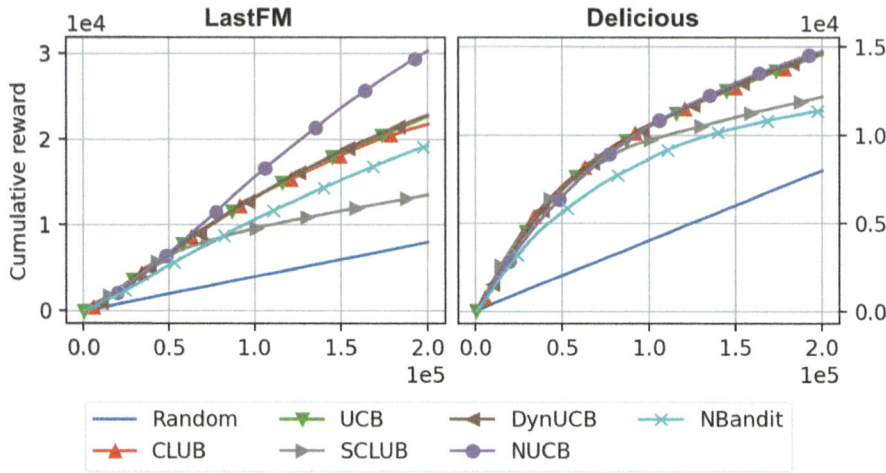

Fig. 2. Cumulative reward of various algorithms for the semi-contextual scenario on LastFM and Delicious.

Figure 1 shows the cumulative CTR (fraction of times the reward is 1 out of the number of retained records so far) of various algorithms on the Yahoo! dataset where we used the replay method.

Figure 2 shows the cumulative reward of various algorithms for the semi-contextual scenario on the LastFM and Delicious datasets. Figure 3 shows the cumulative regret relative to the regret of RANDOM. Regret measures the difference between the rewards obtained by the learner and the rewards obtainable if you know all rewards upfront. We use regret to make comparisons against the experiments in related works easier.

Semi-contextual Scenario. For the semi-contextual scenario, all algorithms struggle to outperform the UCB baseline. On the Yahoo! dataset, CLUB and NUCB score better, but the improvement is minor and not consistent across the entire time horizon. On LastFM only NUCB is able to improve over UCB and on Delicious no algorithms beat this baseline.

For all datasets, the optimal number of clusters for DYNUCB is 1, which means it is essentially identical to UCB. This also indicates that these datasets are not very suitable for clustering. Furthermore, the optimal hyper-parameters for CLUB and SCLUB encourage both algorithms to keep the majority of users in a single cluster. In fact, their performance starts to deteriorate once more users start to split off and form their own clusters. Because of their design, both of these algorithms tend to form one big cluster and many small clusters containing few users. In the semi-contextual scenario, such small clusters are problematic for the same reason why using a separate instance of UCB per user does not work: there are not enough interactions to learn from.

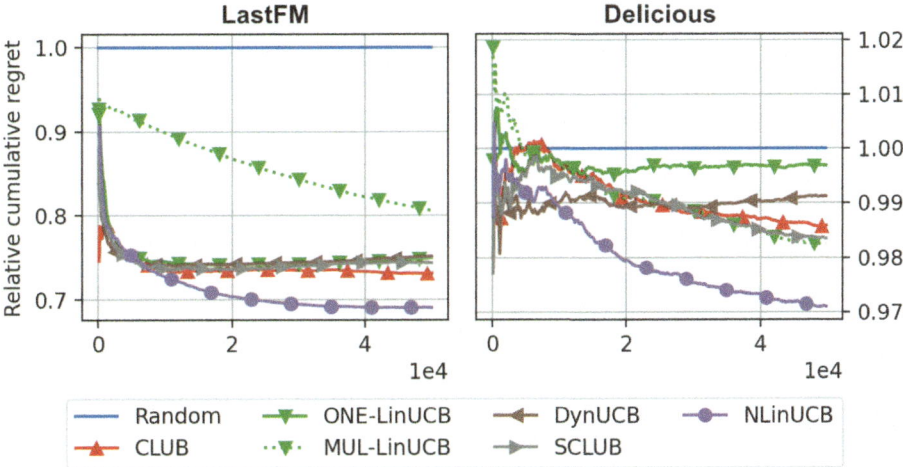

Fig. 3. Cumulative relative regret and cumulative reward of various algorithms for the contextual scenario on LastFM and Delicious.

Contextual Scenario. For the contextual scenario, our algorithm NLINUCB shows improvement over the baselines on both dataset. This is somewhat unexpected, as previous results indicated that Delicious was not very suitable to collaborative filtering because every item is only liked by a few users [9].

Our reproduced results are more or less consistent with related works [9, 10][4] Because the rewards depend less on users for LastFM, it is no surprise that ONE-LINUCB performs better than MUL-LINUCB and the reverse is true for Delicious where personalization is more desired. For delicious, all state-of-the-art clustering based algorithms fall between these two baselines while NLINUCB outperforms them. This indicates that Delicious is more suitable to collaborative filtering than previously thought. On LastFM, CLUB and SCLUB perform better than the LINUCB baselines, but not as much as NLINUCB.

We also observe that in this scenario, CLUB and SCLUB do not suffer when they settle on more clusters as they did for the semi-contextual scenario. With the availability of item features, the aggregated information of a few users in a small cluster can be enough to make good predictions, which is not the case for the semi-contextual scenario.

4.5 Global Information

In our adjusted click and impression count, we explicitly include global information when we add the weighted counts of similar users. This global information

---

[4] On Delicious, the results we report are slightly worse while on LastFM all our results are slightly better. In both cases, the ranking of algorithms remains the same. We attribute this to differences with the tag analysis used to extract item features.

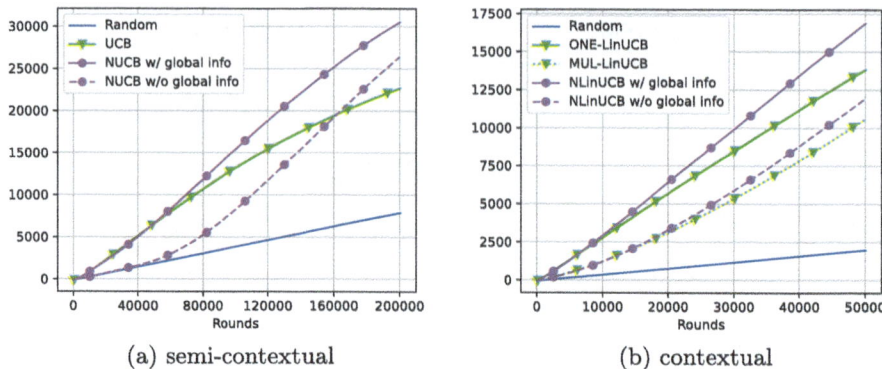

(a) semi-contextual   (b) contextual

Fig. 4. Cumulative reward curves showcasing the effect of global information on lastFM.

turns our to be essential for good performance and the primary reason why our algorithm works well. It allows the agent to keep up with UCB and ONE-LINUCB at the beginning when little similarity information is available, while still enabling personalization during later rounds.

Figure 4 compares the cumulative reward of our algorithm when it does and does not use global information. The results are for LastFM, but the conclusions equally apply to the other datasets. As expected, the version with global information starts out on par with UCB and ONE-LINUCB, and rises above it as time goes on. The variant without global information starts out more comparable with RANDOM or MUL-LINUCB, but does catch up in later rounds.

It should also be noted that the cold start window where little similarity information is available, occurs per user. Thus, in real world scenarios where there are a many new or anonymous users, the advantage of global information applies across all rounds, not only at startup. The global information allows new users to benefit from the wealth of past interaction as if a single instance of UCB was used, while automatically transitioning to a more personalized algorithm as more information becomes available. In this way it enables the best of both worlds.

4.6 User Versus Item-Based Aggregation

Figure 5 compares aggregating based on user or item similarities on LastFM for the semi-contextual scenario. User-based aggregation seems the most effective, but combining both yields the best results.

Figure 6 compares using different values for the similarity weights. As can be seen, the value should be high, and too high generally is not detrimental. The main function of this hyper-parameter is to ensure that the similarity information overpowers the global information after a certain number of iterations and as such the exact value does not matter. This is good news, as it implies our

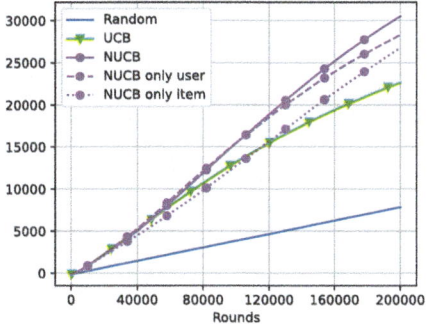

Fig. 5. Cumulative reward curves showcasing the effect of user- versus item-based aggregation for the semi-contextual scenario on LastFM.

proposed method does not have any hyper-parameters of concern besides the base exploration parameter α shared by all UCB-like algorithms.

5   Related Work

The most closely related works are a series of studies on online clustering of bandits [9,10,16–18]. While our work does not rely on clustering, it follows the same general framework of basing predictions on aggregated user information. Our work further makes the same assumptions about what information is available to the learner and that the goal is to achieve high cumulative reward.

Foundational to the work on clustering of bandits is the research on linear bandits [1,3,14]. Among this work, LinUCB [7] is a well known approach. In it's essence these algorithms assume that the rewards linearly depend on the context, which are typically feature vectors associated with users and/or items.

In another line of research, user relationships (e.g. from a social network) are used to speed up contextual bandit learning under the assumption that users share interests with their friends [5,27]. These works are related to ours in their collaborative nature, but different because we learn the user relations online instead of assuming them known apriori. Therefore, they were not included in our comparison.

In another approach to collaborative bandits, the correlation between users and items is captured directly by learning latent features (e.g. through matrix factorization) [6,12,24–26]. These approaches work very differently from those discussed earlier and are mainly motivated by the case where no side information about users or items is available. As our primary focus is on contextual bandits where we do have item features, we do not compare against these methods.

Distinct from other approaches in a recent neighbor-based algorithm which considers the users as the arms instead of the items [19]. While it is also based on neighborhood collaborative filtering, it is fundamentally different because it does exploration of the users instead of the items. It also cannot make use of item features.

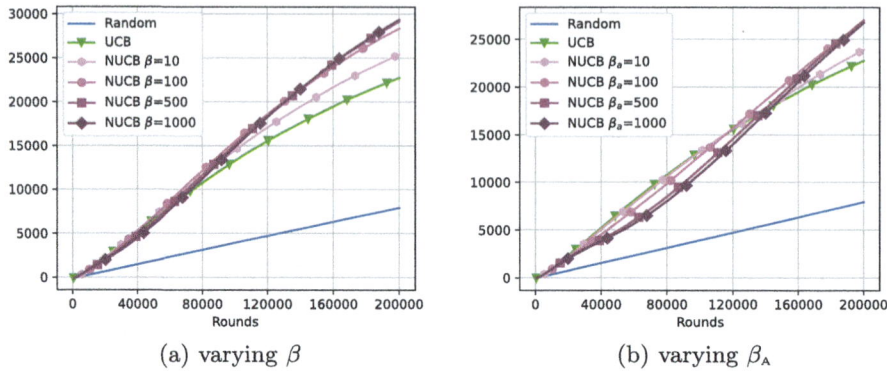

(a) varying β (b) varying β_A

Fig. 6. Cumulative reward curves showcasing the effect of user and item similarity weights for the semi-contextual scenario on LastFM.

6 Conclusion and Future Work

We have developed a novel neighborhood-based bandit algorithm that uses similarity weighting to personalize the predictions of a base bandit algorithm such as UCB or LinUCB. We have demonstrated the effectiveness of our algorithm in a wide range of scenarios by considering multiple datasets and the availability of item features. Importantly, our method incorporates global information to combat data scarcity. This enables our approach to handle cold-start situations effectively, while still offering personalization as more information becomes available. This is in contrast to state-of-the-art clustering bandit algorithms which simply aggregate cluster-level information. When the clusters are small and contain few users, this can lead to poor performance, especially when there are no item features and the item pool is large.

An interesting avenue for future work is to integrate ideas from compression-based approaches such as KRIMP [23] to more efficiently model item histories. Instead of relying on raw click matrices for similarity computation, a compressed representation of user-item interactions could be maintained. This would allow faster and potentially more robust estimation of similarity-weighted click-through rates (CTR), particularly in large-scale settings.

Acknowledgments. This work was funded by the FWO (Research Foundation—Flanders) [11A4D25N to N. Daniëls].

References

1. Abe, N., Long, P.M.: Associative reinforcement learning using linear probabilistic concepts. In: Proceedings of the Sixteenth International Conference on Machine Learning, ICML 1999, pp. 3–11. Morgan Kaufmann Publishers Inc. (1999)
2. Aggarwal, C.C.: Recommender Systems: The Textbook. Springer, Cham (2016). https://doi.org/10.1007/978-3-319-29659-3_2

3. Auer, P.: Using upper confidence bounds for online learning. In: Proceedings 41st Annual Symposium on Foundations of Computer Science, pp. 270–279 (2000). https://doi.org/10.1109/SFCS.2000.892116

4. Cantador, I., Brusilovsky, P., Kuflik, T.: 2nd workshop on information heterogeneity and fusion in recommender systems (hetrec 2011). In: Proceedings of the 5th ACM Conference on Recommender Systems, RecSys 2011. ACM (2011)

5. Cesa-Bianchi, N., Gentile, C., Zappella, G.: A gang of bandits. In: Proceedings of the 26th International Conference on Neural Information Processing Systems, NIPS 2013, vol. 1, pp. 737–745. Curran Associates Inc. (2013)

6. Christakopoulou, K., Banerjee, A.: Learning to interact with users: a collaborative-bandit approach. In: Proceedings of the 2018 SIAM International Conference on Data Mining (SDM), pp. 612–620 (2018). https://doi.org/10.1137/1.9781611975321.69

7. Chu, W., Li, L., Reyzin, L., Schapire, R.: Contextual bandits with linear payoff functions. In: Gordon, G., Dunson, D., Dudík, M. (eds.) Proceedings of the Fourteenth International Conference on Artificial Intelligence and Statistics. Proceedings of Machine Learning Research, vol. 15, pp. 208–214. PMLR (2011)

8. Ferrari Dacrema, M., Cremonesi, P., Jannach, D.: Are we really making much progress? A worrying analysis of recent neural recommendation approaches. In: Proceedings of the 13th ACM Conference on Recommender Systems, RecSys 2019, pp. 101–109. ACM (2019). https://doi.org/10.1145/3298689.3347058

9. Gentile, C., Li, S., Kar, P., Karatzoglou, A., Zappella, G., Etrue, E.: On context-dependent clustering of bandits. In: Precup, D., Teh, Y.W. (eds.) Proceedings of the 34th International Conference on Machine Learning. Proceedings of Machine Learning Research, vol. 70, pp. 1253–1262. PMLR (2017). https://proceedings.mlr.press/v70/gentile17a.html

10. Gentile, C., Li, S., Zappella, G.: Online clustering of bandits. In: Xing, E.P., Jebara, T. (eds.) Proceedings of the 31st International Conference on Machine Learning, pp. 757–765. Proceedings of Machine Learning Research, PMLR (2014). https://proceedings.mlr.press/v32/gentile14.html

11. Guillou, F., Gaudel, R., Preux, P.: Collaborative filtering as a multi-armed bandit. In: NIPS 2015 Workshop: Machine Learning for eCommerce (2015)

12. Kawale, J., Bui, H.H., Kveton, B., Tran-Thanh, L., Chawla, S.: Efficient Thompson sampling for online matrix-factorization recommendation. In: Cortes, C., Lawrence, N., Lee, D., Sugiyama, M., Garnett, R. (eds.) Advances in Neural Information Processing Systems, vol. 28. Curran Associates, Inc. (2015)

13. Koren, Y., Rendle, S., Bell, R.: Advances in collaborative filtering. In: Ricci, F., Rokach, L., Shapira, B. (eds.) Recommender Systems Handbook, pp. 91–142. Springer, Cham (2022). https://doi.org/10.1007/978-1-0716-2197-4_3

14. Li, L., Chu, W., Langford, J., Schapire, R.E.: A contextual-bandit approach to personalized news article recommendation. In: Proceedings of the 19th International Conference on World Wide Web, WWW 2010, pp. 661–670. ACM (2010). https://doi.org/10.1145/1772690.1772758

15. Li, L., Chu, W., Langford, J., Wang, X.: Unbiased offline evaluation of contextual-bandit-based news article recommendation algorithms. In: Proceedings of the Fourth ACM International Conference on Web Search and Data Mining, WSDM 2011, pp. 297–306. ACM (2011). https://doi.org/10.1145/1935826.1935878

16. Li, S., Chen, W., Li, S., Leung, K.S.: Improved algorithm on online clustering of bandits. In: Proceedings of the 28th International Joint Conference on Artificial Intelligence, IJCAI 2019, pp. 2923–2929. AAAI Press (2019)

17. Li, S., Karatzoglou, A., Gentile, C.: Collaborative filtering bandits. In: Proceedings of the 39th International ACM SIGIR Conference on Research and Development in Information Retrieval, SIGIR 2016, pp. 539–548. ACM (2016). https://doi.org/10.1145/2911451.2911548

18. Nguyen, T.T., Lauw, H.W.: Dynamic clustering of contextual multi-armed bandits. In: Proceedings of the 23rd ACM International Conference on Conference on Information and Knowledge Management, CIKM 2014, pp. 1959–1962. ACM (2014). https://doi.org/10.1145/2661829.2662063

19. Sanz-Cruzado, J., Castells, P., López, E.: A simple multi-armed nearest-neighbor bandit for interactive recommendation. In: Proceedings of the 13th ACM Conference on Recommender Systems, RecSys 2019, pp. 358–362. ACM (2019). https://doi.org/10.1145/3298689.3347040

20. Silva, N., Werneck, H., Silva, T., Pereira, A.C., Rocha, L.: Multi-armed bandits in recommendation systems: a survey of the state-of-the-art and future directions. Expert Syst. Appl. **197**, 116669 (2022). https://doi.org/10.1016/j.eswa.2022.116669

21. Sutton, R.S., Barto, A.G.: Reinforcement Learning: An Introduction. MIT Press, Cambridge (2018)

22. Verstrepen, K., Goethals, B.: Unifying nearest neighbors collaborative filtering. In: Proceedings of the 8th ACM Conference on Recommender Systems, RecSys 2014, pp. 177–184. ACM (2014). https://doi.org/10.1145/2645710.2645731

23. Vreeken, J., van Leeuwen, M., Siebes, A.: Krimp: mining itemsets that compress. Data Min. Knowl. Discov. **23**(1), 169–214 (2011). https://doi.org/10.1007/S10618-010-0202-X

24. Wang, H., Wu, Q., Wang, H.: Learning hidden features for contextual bandits. In: Proceedings of the 25th ACM International on Conference on Information and Knowledge Management, CIKM 2016, pp. 1633–1642. ACM (2016). https://doi.org/10.1145/2983323.2983847

25. Wang, H., Wu, Q., Wang, H.: Factorization bandits for interactive recommendation. In: Proceedings of the AAAI Conference on Artificial Intelligence, vol. 31, no. 1 (2017). https://doi.org/10.1609/aaai.v31i1.10936. https://ojs.aaai.org/index.php/AAAI/article/view/10936

26. Wang, Q., Zeng, C., Zhou, W., Li, T., Iyengar, S.S., Shwartz, L., Grabarnik, G.Y.: Online interactive collaborative filtering using multi-armed bandit with dependent arms. IEEE Trans. Knowl. Data Eng. **31**(8), 1569–1580 (2019). https://doi.org/10.1109/TKDE.2018.2866041

27. Wu, Q., Wang, H., Gu, Q., Wang, H.: Contextual bandits in a collaborative environment. In: Proceedings of the 39th International ACM SIGIR Conference on Research and Development in Information Retrieval, SIGIR 2016, pp. 529–538. ACM (2016). https://doi.org/10.1145/2911451.2911528

28. Yang, L., Liu, B., Lin, L., Xia, F., Chen, K., Yang, Q.: Exploring clustering of bandits for online recommendation system. In: Proceedings of the 14th ACM Conference on Recommender Systems, RecSys 2020, pp. 120–129. ACM (2020). https://doi.org/10.1145/3383313.3412250

Interventional idlBNs in DAG-Space

Robert Castelo[(✉)] [iD]

Department of Medicine and Life Sciences, Universitat Pompeu Fabra,
Barcelona, Spain
`robert.castelo@upf.edu`

Abstract. Inclusion-driven structure learning of Bayesian networks, or
idlBNs, converges to the generative structure as the sample size grows
large and as long as that structure is an acyclic digraph (DAG) over
the observed random variables. Because Markov equivalence of Bayesian
networks organizes the search space of DAGs in equivalence classes, an
obvious choice for such an approach is the greedy equivalence search
(GES) algorithm, which carefully traverses the space of essential graphs,
the canonical elements of those equivalence classes, following an inclusion
path. GES is adapted to data produced by multiple intervention experiments in the greedy interventional equivalence search (GIES) algorithm.
The algorithmic complexity of both GES and GIES is in the worst case
exponential in the number of vertices, but it can be reduced to polynomial by bounding the vertex degree during the search, albeit at the cost
of losing the large-sample optimality guarantee. Inclusion-driven structure learning can also be implemented in the search space of DAGs,
as in the hill-climber Monte Carlo (HCMC) algorithm, whose stochastic
nature confers the advantage of a polynomial-time bounded algorithmic complexity. Here, we introduce the interventional HCMC (iHCMC)
algorithm, an inclusion-driven structure learning algorithm for interventional data in DAG-space. Using synthetic Gaussian data, we verify that
iHCMC preserves the large-sample optimality for interventional data
with polynomial-time complexity independent of the sparsity of the generative structure.

Keywords: Bayesian network · inclusion-driven structure learning ·
interventional data

1 Introduction

Graphical Markov models (GMMs) [15,28,44] provide a modular description
of a multivariate distribution by means of labeled graphs without loops and
multiple edges, whose vertices are in one-to-one correspondence with the random variables of that distribution. Different types of graphs determine different
types of GMMs, but they are generally overlapping, i.e., the intersection among
those types of models is nonempty. For instance, graphical models determined
by chordal graphs are at the intersection between those determined by acylic

© The Author(s), under exclusive license to Springer Nature Switzerland AG 2026
M. van Leeuwen and J. Vreeken (Eds.): Arno Siebes Festschrift, LNCS 16067, pp. 161–177, 2026.
https://doi.org/10.1007/978-3-032-03028-3_10

digraphs (DAGs) and unrestricted undirected graphs, and the intersection of models between chordal graphs and transitive DAGs [2] (those where $a \rightarrow b$ and $b \rightarrow c$ imply $a \rightarrow c$) is formed by tree conditional independence (TCI) models that are determined by P_4-free chordal graphs [8,9].

One of the most studied types of GMMs are those determined by DAGs, also known as Bayesian networks, because of their use as a graphical approach to causal inference [33,34,45], and their unique and efficient factorization of a joint multivariate distribution in terms of conditional probabilities or densities of each vertex given its parents. When the DAG structure of the Bayesian network of interest is unknown, one can attempt to learn that structure from available data. Because the number of DAGs grows exponentially in the number of vertices [36], structure learning algorithms use heuristic strategies to traverse that search space [13], potentially getting trapped in local maxima. Alternatively, if the number of vertices is small, one can use dynamic programming approaches that efficiently enumerate, score, and find the global optimum [38]. The search for the Bayesian network that best fits the data can also be biased with informative priors on the network structures [7,31], which can be directly integrated into Markov Chain Monte Carlo (MCMC) procedures when the posterior distribution of Bayesian network structures given the data is of interest [18,29], as for instance in association rule discovery [4] and web mining [17].

Heuristic approaches can be broadly categorized into constraint-based algorithms that use conditional independence hypothesis tests [23,39], score-based algorithms that attempt to maximize some sort of goodness-of-fit score [6,11,13, 25], and hybrid algorithms that combine the previous two strategies [16,41]. In addition to its vast size, a feature of the search space of DAGs that makes the learning problem harder is Markov equivalence, where two different DAGs may represent the same set of conditional independence restrictions [1,10]. One way to approach this complexity is to define the search space in terms of the canonical elements of those equivalence classes, which have been introduced in the literature under different names such as *patterns* [40], completed partially-directed acyclic graphs (*CPDAGs*) [10] and *essential graphs* [1].

Just as we expect a consistent estimator of some quantity of interest to converge in probability to the value that we want to estimate, we may also expect that structure learning algorithms of Bayesian networks have some large-sample optimality guarantees that allow them to converge to the generative structure as the sample size grows large. This is the case of algorithms such as the Peter and Clark (PC) algorithm [23,39], the greedy equivalence search (GES) [11], or the hill-climber Monte Carlo (HCMC) [6,25], under the assumption that the generative structure is a DAG over the observed variables. The PC algorithm is constraint-based, while GES and HCMC are score-based.

However, these algorithms preserve this optimality only for independent and identically distributed (iid) observational data. In a causal inference setting, where researchers conduct experiments with *interventions* that set one or more variables to specific values that remove their original causal dependencies, the resulting data points may be independent, but not identically distributed. Learning the structure of Bayesian networks from a mixture of observational and interventional data has been addressed using a simple greedy search algorithm [14] and adapting GES to this setting in the so-called greedy interventional equivalence search (GIES) algorithm [20].

Due to the way in which graphical manipulations have to be performed to traverse the space of essential graphs, the algorithmic complexity of both GES and GIES is exponential in its worst case [11,20]. This also happens with the PC algorithm due to the way hypothesis tests are performed [23]. However, if the underlying generative structure is sparse, which is a reasonable assumption in most real-world applications, GES, GIES and PC run in polynomial time. Alternatively, PC, GES and GIES may be run considering a bound on the maximum vertex degree in the graph, which leads to a polynomial-time complexity algorithm, albeit at the cost of losing the large-sample optimality guarantee. On the other hand, HCMC is bounded by polynomial time because of the stochastic nature of its search strategy, independently of the sparsity of the generative structure.

In this paper, we introduce a version of the HCMC algorithm adapted to data from experiments with multiple interventions. Simulating synthetic Gaussian data, we show that just as GIES, this *interventional* HCMC algorithm, which we shall call iHCMC hereafter, preserves its large-sample optimality guarantee with interventional data, while running in polynomial time without the need to restrict the vertex degree in the search space.

The rest of this chapter is organized as follows. In the next section, we introduce background concepts, terminology, and notation on GMMs. In Sects. 3 and 4, we briefly review, respectively, inclusion-driven structure learning with the GES and HCMC algorithms, and structure learning from interventional data with the GIES algorithm. In Sect. 5, we describe the iHCMC algorithm. In Sect. 6 we empirically show using simulated data the optimal properties of iHCMC. Finally, we conclude this chapter with a discussion in Sect. 7.

2   Background Concepts, Terminology and Notation

The background concepts, terminology, and notation introduced here have been borrowed from the books of Harary [19], Whitakker [44], Lauritzen [28], and Cox and Wermuth [15]. The reader may consult these books for a more comprehensive account of graphs and GMMs.

2.1   Graphs, Paths and Separation

A *graph* is a pair $G = (V, E)$ where $V = \{1, \ldots, p\}$ is a finite set of p vertices and $E \subseteq V \times V$ is a subset of edges. Here we consider only graphs that are labeled, where all vertices are distinct, and simple, that is, without loops (edges whose endpoints are the same vertex) and multiple edges (more than one edge between two endpoints). Two vertices $u, v \in V$ are *adjacent* in G if $\{u, v\} \in E$, denoted by $u \sim b$.

A *walk* of length $k \geq 2$ between two vertices x and y in a graph G is a sequence $\pi_{xy} = \langle v_1 = x, \ldots, v_k = y \rangle$ such that $\{v_i, v_{i+1}\} \in E$ for every $i = 1, \ldots, k-1$. A *trail* is walk in which all edges $\{v_i, v_{i+1}\} \in E$ are distinct. A *path* is a trail in which all vertices (and therefore also all edges) are distinct. A *circuit* is a nonempty trail whose endpoints are the same vertex. A *cycle* is a circuit in which only the first and last vertices are the same vertex, i.e., $\pi_{xx} = \langle v_1 = x, \ldots, v_k = x \rangle$.

An edge (u, v) is *directed*, also known as an *arc*, if and only if $(u, v) \in E$, but $(v, u) \notin E$. A directed edge between two vertices u and v is represented graphically by an arrow pointing from u toward v, i.e., $u \rightarrow v$. A graph $G = (V, E)$ is directed if all edges in E are directed edges. In a directed edge $u \rightarrow v$, we call u the *parent* of v and v the *child* of u. All vertices in a directed graph with a common child vertex v are known as the *parent set* of v, denoted by $pa(v)$.

A *directed path* $\pi_{xy} = \langle v_1 = x, \ldots, v_k = y \rangle$ is a direction-preserving path, where every edge on the path points in the same direction, i.e., $v_i \rightarrow v_{i+1}$, while a directed cycle is a direction-preserving cycle. An *acyclic directed graph* $G = (V, E)$, popularly known as a DAG, is a directed graph without directed cycles. The *skeleton* of a DAG is the undirected graph obtained by replacing the directed edges by undirected ones, preserving the adjacencies.

We say that a triplet of vertices (x, y, z) in a DAG $G = (V, E)$ forms a *V-configuration* if G has a skeleton such that $x \sim y$ and $y \sim z$, but $x \nsim z$. We shall distinguish between the following three types of V-configurations: *transition-oriented* where $x \rightarrow y \rightarrow z$, *source-oriented* where $x \leftarrow y \rightarrow z$, and *sink-oriented* where $x \rightarrow y \leftarrow z$. The y vertex in a sink-oriented V-configuration is commonly known as a *collider* vertex.

Given three disjoint subsets of vertices $A, B, S \subset V$, where A and B are nonempty, we can classify every path between the vertices in A and B into *active* or *blocked*, according to the membership in S of collider and noncollider vertices of that path. When all paths between vertices in A and B are blocked by S, we say that S *separates* A from B in G and we will denote it by $A \perp_G B | S$; see [15, 28, 44] for full details on the concept of separation in DAGs, commonly known as *d-separation*. An important intuition behind this and any other definition of separation in graphs is that no matter what subset $S \subset V$ we consider, two vertices $x, y \in V$ in a DAG $G = (V, E)$ directly connected by an edge, i.e., $\{x, y\} \in E$, cannot be separated by S. Therefore, a necessary condition for two vertices x, y to be separated is that they are not adjacent, i.e., $\{x, y\} \notin E$.

2.2   Markov Properties, Equivalence and Inclusion

Let $X \equiv X_V$ be a random vector indexed by a finite set $V = \{1, \ldots, p\}$ with a joint multivariate distribution $P_V \equiv P(X_V)$. Let $G = (V, E)$ be a DAG whose vertices are in one-to-one correspondence with the random variables in X_V. Given three disjoint subsets of random variables $X_A, X_B, X_S \subset X_V$, where X_A and X_B are nonempty, we say that X_A is conditional independent (CI) of X_B given X_S in P_V, and denote it by $X_A \perp\!\!\!\perp X_B | X_S$, or $A \perp\!\!\!\perp B | S$ for short, if and only if the joint density factorizes using the margins defined by A, B and S as, for instance, in $f_{ABS}(a, b, s) = f_{AS}(a, s)f_{BS}(b, s)/f_S(s)$ for all s with $f_S(s) > 0$.

Using d-separation relationships $\perp_G$, we can encode or represent, in a DAG G, a subset of the CI restrictions that hold in P_V. We say that P_V obeys the *global Markov property* relative to a DAG G, or that P_V is *Markov over a DAG G*, if for every triplet of disjoint subsets $A, B, S \subset V$, where A and B are nonempty, $A\perp_G B|S \Rightarrow A\perp\!\!\!\perp B|S$ in P_V. Additional Markov properties exist using graphical criteria other than d-separation; see [28] for further details.

We call the family of all joint multivariate distributions P_V Markov over a DAG G, denoted by $\mathcal{M}(G)$, the GMM determined by G, or to simplify terminology and notation, the Bayesian network G. A multivariate density function $f(x)$ for $P_V \in \mathcal{M}(G)$ also obeys the global Markov property relative to a DAG G, if it admits the following unique factorization in terms of conditional densities of each random variable given its parents in G [28]:

$$f_G(x) = \prod_{i \in V} f_G(x_i|x_{pa(i)}). \tag{1}$$

A distinctive feature of Bayesian networks, with respect to some other types of GMMs such as those determined by undirected graphs, is that two different Bayesian networks G and G' may represent the same set of CI restrictions as in, e.g., $a \leftarrow b \leftarrow c$, $a \leftarrow b \rightarrow c$ and $a \rightarrow b \rightarrow c$, which all represent $X_a \perp\!\!\!\perp X_c|X_b$. In such a case, $\mathcal{M}(G) = \mathcal{M}(G')$, and one says that G and G' are *Markov equivalent*. From a graphical perspective, two different DAGs G and G' are Markov equivalent if and only if they have the same skeleton and the same sink-oriented V-configurations [43].

The canonical element of a Markov equivalence class of DAGs is represented by an essential graph (EG) [1], formed by directed and undirected edges without partially directed cycles. In an EG, an edge is directed if an only if that edge has the same orientation in all DAGs from the equivalence class it represents, otherwise it is undirected. Given a DAG G in an equivalence class with two or more members, G should have at least one arc that can be reversed in such a way that the resulting DAG remains in the same equivalence class. Such arcs are said to be *covered* and are characterized as follows [10].

Definition 1 *(Covered arc). Given a DAG $G = (V, E)$, an arc $a \rightarrow b$ is covered in G if $pa(b) = \{a\} \cup pa(a)$.*

The previous definition means that an arc $a \rightarrow b$ is covered if and only if the parent sets of a and b are identical except for the a vertex, which is a parent of b, but it cannot be a parent of itself. Covered arcs provide the following additional characterization of Markov equivalence.

Lemma 1 *(Lemma 2.1 [6]). Given two different Markov equivalent DAGs G and G', there exists a sequence $L_1, \ldots, L_n$ of DAGs such that $L_1 = G$ and $L_n = G'$ and L_{i+1} is obtained from L_i by reversing a covered edge in L_i, for $i = 1, \ldots, n - 1$.*

From the previous definitions of Markov equivalence, it follows that two DAGs G and G' with a different number of arcs cannot be Markov equivalent. Two extreme such cases are the complete DAG G_c, where all vertices are adjacent and no CI restriction can be represented, and the empty DAG $G_\emptyset$, where no arc is present and, therefore, all possible CI restrictions are represented. Clearly, the family of all P_V Markov over $G_\emptyset$ is more constrained than the one that is Markov over G_c, and therefore we may say that $\mathcal{M}(G_\emptyset)$ is included in $\mathcal{M}(G_c)$, i.e., $\mathcal{M}(G_\emptyset) \subset \mathcal{M}(G_c)$ [6,25,26]. We call this relationship a *Markov inclusion order*, and can also be defined using only subsets of CI restrictions as follows. Given a DAG $G = (V, E)$,

$$\mathcal{M}'(G) = \{(A, B, S) : A, B \neq \emptyset \wedge A \perp_G B | S\}, \tag{2}$$

defines the set of CI restrictions that could be read off the DAG G. Under this definition, $\mathcal{M}'(G_c) \subset \mathcal{M}'(G_\emptyset)$. The collection of subsets of CI restrictions for all DAGs on p vertices $\{\mathcal{M}'(G_1), \ldots, \mathcal{M}'(G_k)\}$ forms a partial ordered set, or *poset*, with a partial order relation defined by Markov inclusion [6, Fig. 2]. A sequence of DAGs $G_1, \ldots, G_k$ forms an *inclusion path* if either $\mathcal{M}'(G_1) \subseteq \mathcal{M}'(G_2) \subseteq \cdots \subseteq \mathcal{M}'(G_k)$ or $\mathcal{M}'(G_1) \supseteq \cdots \supseteq \mathcal{M}'(G_{k-1}) \supseteq \mathcal{M}'(G_k)$.

Given two DAGs G and G' that are not Markov equivalent, deciding whether $\mathcal{M}'(G) \subset \mathcal{M}'(G')$, $\mathcal{M}'(G) \supset \mathcal{M}'(G')$ or whether G and G' are not in inclusion, was an open problem for a long time. In 1988, Verma and Pearl [42] attempted to give necessary and sufficient conditions to characterize the inclusion order of Bayesian networks. Later in 1997, Chris Meek in his PhD thesis [30] provided the following conjecture on the inclusion problem.

Conjecture 1 (Meek's conjecture [30]). Given two Bayesian networks G and G', $\mathcal{M}'(G) \subseteq \mathcal{M}'(G')$ if and only if there exists a sequence of DAGs $G_1, \ldots, G_n$ such that $G_1 = G$, $G_n = G'$ and the G_{i+1} is obtained from G_i by either reversing a covered arc or removing an arc, for $i = 1, \ldots, n - 1$.

In 2001, Kočka, Bouckaert and Studený [26] showed that the conditions given by Verma and Pearl in [42] were necessary but not sufficient, and gave a proof to Meek's conjecture for the particular case in which G and G' differ in at most one adjacency. Finally, in 2002, Max Chickering [11] provided a general proof of Meek's conjecture by means of the following theorem.

Theorem 1 *(Theorem 4 [11]). Let G and G' be any pair of DAGs such that $\mathcal{M}'(G) \subseteq \mathcal{M}'(G')$. Let r be the number of arcs in G that have opposite orientation in G', and let m be the number of arcs in G' that do not exist in either orientation in G. There exists a sequence of at most $r + 2m$ distinct arc reversals and additions in G' with the following properties:*

1. *Each arc reversed is a covered arc.*
2. *After each reversal and addition, G' is a DAG and $\mathcal{M}'(G) \subseteq \mathcal{M}'(G')$.*
3. *After all reversals and additions $G = G'$.*

As we shall see in the next section, the Markov inclusion order is relevant for structure learning of Bayesian networks, because it enables building learning algorithms with a large-sample optimality guarantee.

3 Inclusion-Driven Structure Learning

Score-based algorithms for learning the structure of a Bayesian network from data consist of a scoring metric and a search strategy. Several scoring metrics have been introduced for discrete and continuous data, e.g., [3,13,22,27,37], based on the unique factorization of the probability mass or density function that one may derive from a given DAG structure (see Eq. 1 in the previous section). A useful property of some of these score metrics, such as BDe [22], BGe [27] or BIC [37], is local consistency. A score metric is *locally consistent* [6,11] if, given data sampled from a joint multivariate distribution P_V, it increases after adding an arc that removes a CI restriction that does not hold in P_V, and it decreases after adding an arc that removes a CI restriction that holds in P_V, i.e., it will increase when an *unnecessary* arc is removed.

A straightforward search strategy is a greedy hill-climbing algorithm that starts from a DAG G, typically the empty DAG without edges $G_\emptyset$, i.e., $G = G_\emptyset$, scores it, generates all possible neighboring DAGs with one more arc, scores them, and selects the one with the highest score; let us call it G'. If the score for G' is higher than the score for G, then $G = G'$, we start again generating all possible neighboring DAGs and continue in this loop until the score does not improve. When the algorithm stops, G is the DAG with the highest score found by the algorithm.

The set of neighboring DAGs can be generated in different ways, where a simple one is to create all DAGs derived from adding one arc in each empty adjacency, as long as it does not introduce a directed cycle, removing every present arc, and reversing every present arc, as long as it does not introduce a directed cycle. It has been shown [6,11] that this simple hill-climbing algorithm gets easily stuck in local maxima, even when using a locally consistent scoring metric.

However, if the search strategy is such that the neighboring DAGs generated at each step enable following an inclusion path, then in the limit of the size of the sample, the hill-climbing algorithm will converge to the generative structure [6,11,12,25,26]. This is the case for the hill-climber Monte carlo (HCMC) algorithm [6,25] that works in the space of DAGs, and the greedy equivalence search (GES) algorithm that works in the space of EGs [11]. A distinctive feature of the HCMC algorithm is that it is stochastic, which enables following an inclusion path with positive probability, while bounding its algorithmic complexity to polynomial time. On the other hand, GES deterministically attempts to follow an inclusion path, but the graphical operations required to generate neighboring EGs have a worst-case exponential algorithmic complexity.

Structure learning approaches other than the previous greedy strategies can also lead to optimal algorithms, as long as they attempt to follow an inclusion path. This is the case of the constraint-based PC algorithm [39] or the MCMC-based eMC^3 algorithm [6,25].

4   Interventional Structure Learning

Because two different Markov equivalent DAGs G and G' also lead to two equivalent factorizations $f_G(x) = f_{G'}(x)$, factorizing a joint multivariate distribution according to a DAG is necessary but not sufficient to make a causal interpretation of that DAG. To approach such a causal interpretation, we may consider *interventional probability distributions* using the *do* operator [35, pg. 70]:

$$f_G(Y = y|do(X = x)),\tag{3}$$

by which we obtain the density of $Y = y$ when we intervene in X, by setting the value $X = x$. Given a DAG $G = (V, E)$, we call the *intervention targets* [20] the subset of vertices $I \subseteq V$ indexing the variables $X_I \subseteq X_V$ that we want to intervene in. When we *intervene* in one variable with $do(X = x)$, we will assume that such intervention is *modular* [35, pg. 63], i.e., changing the causal mechanism in one of the variables in the system does not change the causal mechanism in other variables. More intuitively, modularity implies that the intervention only affects the incoming edges of the intervened variable by removing them, while the rest of the graph structure remains intact. This notion can be formalized in the following definition of an *intervention graph* [20, Definition 5].

Definition 2 *(Intervention graph). Let $G = (V, E)$ be a DAG with vertex set V and edge set E, and $I \subseteq V$ a subset of intervention targets. The intervention graph of G is the DAG $G^{(I)} = (V, E^{(I)})$, where $E^{(I)} := \{(a, b)|(a, b) \in E, b \notin I\}$.*

Modularity also implies that if we intervene in a subset of variables $X_I \subseteq X_V$, when $i \notin I$ then $f_G(x_i|x_{pa(i)})$ remains unchanged, while if $i \in I$ then $f_G(x_i|x_{pa(i)}) > 0$ only if $do(X_i = x_i)$ and $f_G(x_i'|x_{pa(i)}) = 0$ for every other $x_i' \neq x_i$. Consequently, altering the graph structure after an intervention leads to the following *truncated factorization* [35, pg. 24] [20]:

$$f_G(X_V = x_V|do(X_I = x_I)) = \prod_{i \notin I} f(x_i|x_{pa(i)}) \prod_{i \in I} \tilde{f}(x_i),\tag{4}$$

where $\tilde{f}(x_i)$ denotes the joint product density for $do(X_i = x_i)$. When $I = \emptyset$, then $f_G(X_V = x_v|do(X_I = x_I)) = f_G(X_V)$, including the observational density as a specific case of a truncated factorization without interventions.

A *family of targets* $\mathcal{I} = \{I_j\}_{j=1}^J$ [20, pg. 2412] is a collection of subsets of targets. Without loss of generality, assuming that I is either the empty set or a single-vertex intervention target, an example of a family of targets could be $\mathcal{I} = \{\emptyset, \{1\}, \{2\}, \{3\}\}$, for a DAG G with at least 3 vertices.

Given a family of targets $\mathcal{I}$ and a DAG G, an interventional Bayesian network, or simply an interventional DAG, is a family of joint multivariate distributions $\mathcal{M}_{\mathcal{I}}(G)$ Markov over G and therefore $\mathcal{M}_{\emptyset}(G) = \mathcal{M}(G)$. Two different DAGs G and G' are $\mathcal{I}$-Markov equivalent when $\mathcal{M}_{\mathcal{I}}(G) = \mathcal{M}_{\mathcal{I}}(G')$. In graphical terms, this implies that for every $I \in \mathcal{I}$, the two intervention graphs $G^{(I)}$ and $G'^{(I)}$ have the same skeleton and the same sink-oriented V-configurations [20]. This definition of Markov equivalence for interventional DAGs provides a finer partition of Markov equivalence classes.

An *interventional data set* D of sample size n generated by one or more intervention graphs $G^{(I_1)}, \ldots, G^{(I_k)}$ through a family of targets $\mathcal{I} = \{I_1, \ldots, I_k\}$, is the matrix of values $D = \{x_{ij}\}_{n \times p}$, where x_i. are independent but not identically distributed data points sampled from different interventions, specified for every row of that matrix by the *target index vector* $\mathcal{T} = (T^{(1)}, \ldots, T^{(n)})$ with $T^{(i)} = I_l$ and $I_l \in \mathcal{I}$, assuming that each target $I \in \mathcal{I}$ occurs at least once in $\mathcal{T}$. When at least one $T^{(i)} = \emptyset$ and one $T^{(j)} \neq \emptyset$, then D is contains a mixture of observational and interventional data.

Given a data set D of multivariate Gaussian data, resulting from a mixture of observational and interventional data, specified by a family of targets $\mathcal{I}$ and a target index vector $\mathcal{T}$, a Bayesian information criterion (BIC) score for this type of data was introduced in [20] and shown to be *consistent* in [21], i.e., in the limit $n \to \infty$ of the sample size of D, the DAG G maximizing the BIC is the generative structure G from which D was sampled.

These results allowed Hauser and Bühlmann in [20] to derive a generalization of the GES algorithm for the structure learning of Bayesian networks from interventional data, in what they called the Greedy Interventional Equivalence Search (GIES) algorithm.

5   The Interventional HCMC Algorithm

Here, we adapt HCMC to interventional data. The concept of Markov equivalence for interventional DAGs leads to a new, more restrictive concept of a covered arc, the *interventional covered arc*, or $\mathcal{I}$-covered arc for short.

Definition 3 *($\mathcal{I}$-covered arc). Given a DAG $G = (V, E)$ and a family of targets $\mathcal{I}$, $a \to b$ is $\mathcal{I}$-covered in G if $pa(b) = \{a\} \cup pa(a)$ and $I \cap \{a, b\} = \emptyset$ for all $I \in \mathcal{I}$.*

We can now write the following interventional repeated covered arc reversal (iRCAR) algorithm, whose input is a DAG, a parameter r of the maximum number of $\mathcal{I}$-covered arc reversals, and a family of targets $\mathcal{I}$.

```
begin algorithm ircar(dag, r, targets)
01        rr ← rnd(0, r)
02        for i ← 0 to rr do
03                cov_arcs ← covered_arcs(dag)
04                mask ← cov_arcs in targets
05                i_cov_arcs ← cov_arcs[not mask]
06                j ← rnd(0, length(i_cov_arcs)-1)
07                dag ← reverse_arc(dag, i_cov_arcs[j])
08        endfor
09        return dag
endalgorithm
```

In this algorithm, the function `covered_arcs()` returns a vector of covered arcs from the input DAG, while the function `reverse_arc()` returns the DAG given in the first argument, reversing the arc specified in the second argument.

Because the HCMC algorithm works in DAG-space, we can readily use the BIC score for interventional data described in [20,21]. Using this BIC score and the previous iRCAR algorithm we can generalize as follows the HCMC algorithm to interventional data, in what we shall call the *interventional HCMC* algorithm, or iHCMC for short.

```
begin algorithm ihcmc(data, r, maxtrials, targets, target_index)
01      dag ← emptydag(data)
02      trials ← 0
03      local_maximum ← false
04      while not local_maximum do
05            dag1 ← ircar(dag, r, targets)
06            ne_dags ← nicr(dag1, targets)
07            dag1 ← argmax(score(data, ne_dags,
08                                    targets, target_index))
09            s ← score(data, dag, targets, target_index)
10            s1 ← score(data, dag1, targets, target_index)
11            local_maximum ← s1 <= s
12            if not local_maximum then
13                  dag ← dag1
14                  trials ← 0
15            else if trials < maxtrials then
16                  dag ← ircar(dag, r, targets)
17                  local_maximum ← false
18                  trials ← trials + 1
19            endif
20      endwhile
21      return dag
endalgorithm
```

The input parameters for the iHCMC algorithm are a data set, a parameter r of the maximum of $\mathcal{I}$-covered arc reversals, a maximum number of trials escaping from local maxima, a family of targets $\mathcal{I}$ and a target index vector $\mathcal{T}$. The `nicr()` function, given an input DAG, generates all neighboring DAGs with one arc added, one arc removed, and every non-$\mathcal{I}$-covered arc reversed. The `score()` function implements the BIC score function for interventional data described in Sect. 4. The `argmax()` function returns the DAG with the highest BIC score.

6   Experimental Evaluation

In this section, we show the results of assessing the performance of iHCMC and other competing algorithms in learning the structure of Bayesian networks from mixtures of simulated observational and interventional data. We have replicated the experimental evaluation strategy used in the assessment of the GIES algorithm [20] with a few modifications that we will specify where they apply.

We have used the R package `pcalg` [24] to simulate Bayesian networks, Gaussian data from them, and run the algorithms we compare with iHCMC, for which we have developed our own implementation in R. As in [20], we use as a baseline comparison a learning method that does not have a large-sample optimality guarantee, the greedy DAG search (GDS) algorithm that operates in DAG-space by simply adding, removing or reversing arcs, and which is implemented in the function `gds()` of the `pcalg` package. From this package, we have also used the functions `ges()` and `gies()` to run, respectively, the GES and GIES algorithms. We have used the BIC score for Gaussian observational and interventional data described in [20] and implemented in the `pcalg` package through the object classes `GaussLOpenObsScore` and `GaussLOpenIntScore`, respectively.

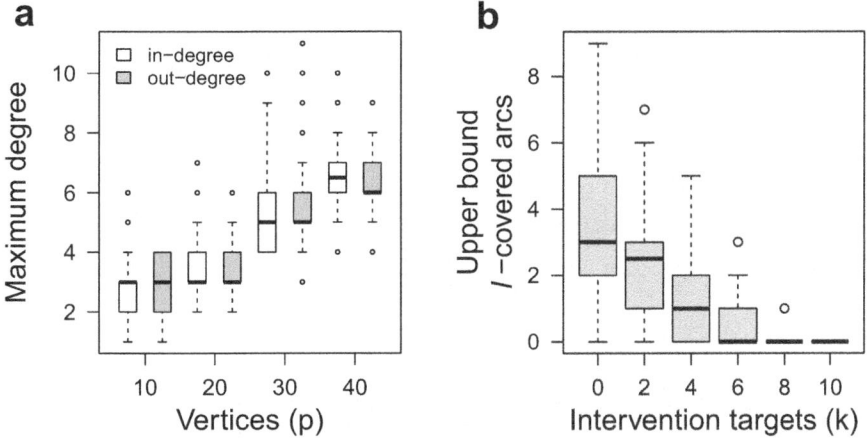

Fig. 1. Simulated intervention DAGs. In (a) maximum in- and out-degree across 100 DAGs for four different combinations of dimension and sparsity rate. In (b) upper bound of the number of $\mathcal{I}$-covered edges as a function of the number k of intervention vertices for 100 DAGs of $p = 10$ vertices.

To simulate random Gaussian Bayesian networks, their parameters and multivariate observational and interventional data from them, we have used methods described in [20] and implemented in the functions `r.gauss.pardag()` and `rmnorm.ivent()` of the `pcalg` package. Unlike the evaluation of the GIES algorithm in [20], where they ran the GDS algorithm with a BIC scoring function only for interventional data, we have also run GDS with a scoring function for observational data. To distinguish between these two regimes of the GDS algorithm, we have labeled the observational data regime with `GDS`, and the interventional one with `iGDS`. We have used the following parameters in our simulations.

– Four different combinations of DAG dimension and sparsity rate $(p, s) \in \{(10, 0.2), (20, 0.1), (30, 0.1), (40, 0.1)\}$ and for each of them we have simulated 100 random DAGs.
– Families of intervention targets of the form $\mathcal{I} = \{\emptyset, I_1, \ldots, I_k\}$, where $I_1, \ldots, I_k$ are k different, randomly chosen intervention targets of size 1 and $k = \{0, 0.2p, 0.4p, 0.6p, 0.8p, p\}$. When $k = 0$, i.e., without intervention targets, the simulated data is purely observational.
– Eight increasing sample sizes $n \in \{50, 100, 200, 500, 1000, 2000, 5000, 10000\}$.

Figure 1b shows the upper bound of the number of $\mathcal{I}$-covered edges as a function of the number k of intervention targets of size 1, across the 100 randomly generated DAGs. It partly reproduces Fig. 8 in [20] with 100 instead of 1000 DAGs, because the number of non-$\mathcal{I}$-essential arrows reported in that figure is an upper bound for the number of $\mathcal{I}$-covered edges at any given stage of the iHCMC algorithm.

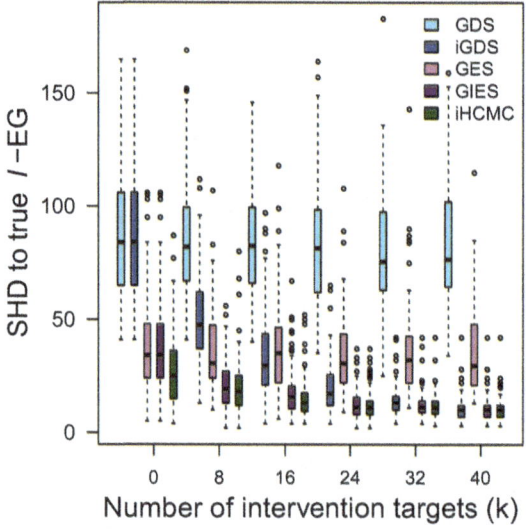

Fig. 2. Structural hamming distance (SHD) between estimated and true $\mathcal{I}$-EGs as function of the number k of single-vertex intervention targets, across 100 random DAGs generated of $p = 40$ vertices. Each data set contains $n = 1000$ data points.

As a measure of divergence between the underlying simulated DAG and the one estimated with one of the learning algorithms, we use the structural Hamming distance (SHD) implemented in the function shd() of the pcalg package [23]. We ran GDS, iGDS, GES, GIES and iHCMC across the data sets simulated from the different combinations of the parameters given before, and calculated the SHD with respect to the generative $\mathcal{I}$-essential graph ($\mathcal{I}$-EG). In the case of the GDS, iGDS and iHCMC algorithms, we first transform the resulting DAG into its corresponding $\mathcal{I}$-EG, and then calculate the SHD.

We ran iHCMC with arguments r=20 and maxtrials=5 throughout all simulations. Figure 2 shows for $p = 40$ how the SHD decreases for GIES, iGDS, and iHCMC as the number of single-vertex intervention targets increases, while the SHD does not improve for GDS and GES, which are only designed to work with observational data. Similar results are obtained with $p = \{10, 20, 30\}$.

As observed in [20], GES and GIES, and GDS and iGDS in our simulations, perform identically with observational data $(k = 0)$, while iGDS, GIES, and iHCMC, perform similarly when the number of single-vertex intervention targets exceeds 50% of the vertices. Likewise, the similar performance of iGDS to GIES and iHCMC is due to the finer partition of interventional Markov equivalence classes, which become singletons when $k = p$ because every DAG becomes its own $\mathcal{I}$-EG. The performance of iGDS underscores the importance of using a scoring function that takes into account the intervention information at each data point.

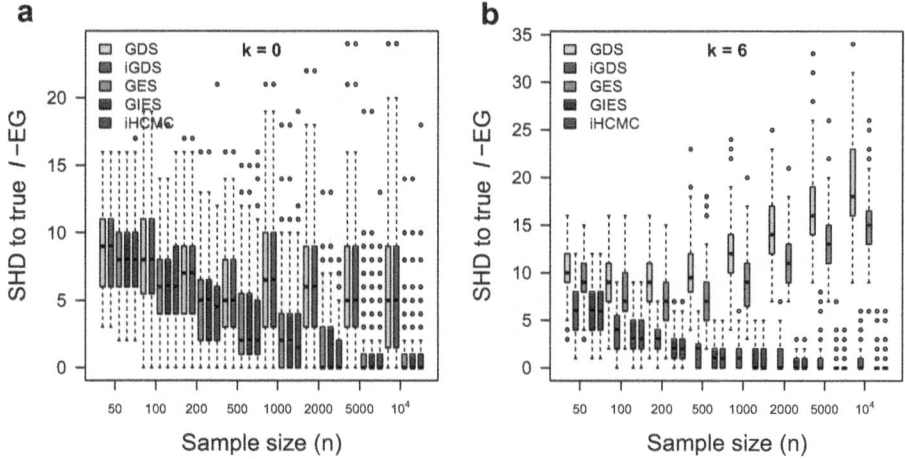

Fig. 3. Structural hamming distance (SHD) between estimated and true $\mathcal{I}$-EGs as function of the sample size, for two different number of intervention targets of size $m = 1$ on DAGs with $p = 10$ vertices, $k = 0$ (a) and $k = 6$ (b).

Figure 3 shows the convergence of the different algorithms to the generative structure, as the sample size grows large, in terms of SHD, for two different numbers of single-vertex intervention targets $(k = \{0, 6\})$, where $k = 0$ corresponds to observational data. We may see that with observational data $(k = 0)$, GES, GIES, and iHCMC converge to the generative structure in the limit of the size of the sample, while this does not happen with GDS or iGDS, since these are not inclusion-driven algorithms and therefore have no large-sample optimality guarantees. With interventional data, GDS and GES learn structures that diverge as the sample size and the number of intervention targets increases.

We simulated DAGs with four different dimensions and sparsity rates, which may result in different degrees of connectivity among the vertices. Sparsity and, in particular, the underlying vertex degree should be taken into account if we are going to restrict the maximum number of edges per vertex in the learning algorithm. Figure 1a shows the maximum in- and out-degree of the 100 simulated DAGs for each of the four different dimensions. We can see that for these simulated DAGs, the maximum vertex degree increases with the dimension.

We have assessed the influence of restricting the vertex degree in the search space by running the algorithms again on the data simulated from DAGs with $p = 10$ vertices and setting GES and GIES to explore the search space with a maximum vertex degree of 3, which should be sufficient for a majority of the underlying DAGs, as shown in Fig. 1a. The results in Fig. 4 reveal that GES (for $k = 0$) and GIES lose their optimality observed in the previous simulation in Fig. 3.

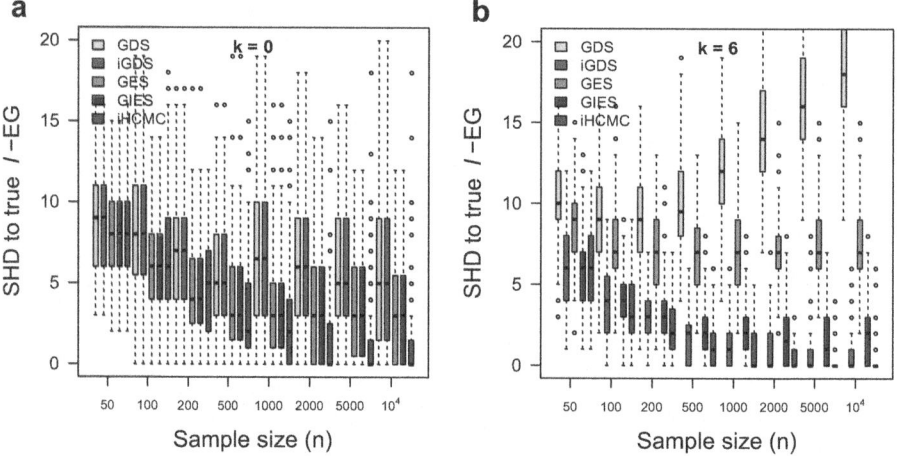

Fig. 4. Structural hamming distance (SHD) between estimated and true $\mathcal{I}$-EGs as function of the sample size, for two different number of intervention targets of size $m = 1$ on DAGs with $p = 10$ vertices, $k = 0$ (a) and $k = 6$ (b). Here, GES and GIES were run with a maximum vertex degree of 3 in the estimated graph.

7   Discussion

Inclusion-driven learned Bayesian networks, or idlBNs[1], have a large-sample optimality guarantee that becomes useful in applications where the available sample size is high, such as in the prediction of spliceosome binding sites on DNA sequences [5]. The trade-off between greediness and randomness that the HCMC

---

[1] Pronounced *ideal BNs*.

algorithm provides has been shown [32] to be also a useful feature when learning from data with a large number of local optima.

Data from intervention experiments, such as in clinical trials, molecular manipulations, or public policy, convey information and meet assumptions that can be harnessed to perform causal inference. Adapting learning algorithms to interventional data contributes to exploiting the results produced by often non-trivial and expensive experiments.

Here we have adapted HCMC, an inclusion-driven structure learning algorithm of Bayesian networks in DAG-space [6,25], to interventional data in the iHCMC algorithm, building on the previous work that led to the development of the GIES algorithm, which performs the same task in EG-space.

We have empirically verified using simulated synthetic data with interventions that iHCMC preserves the same optimal properties as the GIES algorithm and has the advantage that it does not require to bound the maximum vertex degree in the search space to keep the algorithmic complexity in polynomial time.

Acknowledgments. This work was supported by the research project PID2019-105595GB-I00 funded by the MICIU/AEI/10.13039/501100011033. The author thanks the anonymous reviewers for useful comments and suggestions that have helped improve this chapter and the developers and contributors of the R package `pcalg`, which has greatly facilitated running the experimental simulations. The author also wishes to express his most sincere gratitude to Arno Siebes for his mentorship, guidance, and support throughout the pre-doctoral training and beyond.

Disclosure of Interests. The author has no competing interests to declare that are relevant to the content of this article.

References

1. Andersson, S., Madigan, D., Perlman, M.: A characterization of Markov equivalence classes for acyclic digraphs. Ann. Stat. **25**, 505–541 (1997)
2. Andersson, S.A., Madigan, D., Perlman, M.D., Triggs, C.M.: On the relation between conditional independence models determined by finite distributive lattices and by directed acyclic graphs. J. Stat. Plann. Inference **48**(1), 25–46 (1995)
3. Buntine, W.: Theory refinement on Bayesian networks. In: D'Ambrosio, B.D., Smets, P., Bonissone, P.P. (eds.) Proceedings of the Conference on Uncertainty in Artificial Intelligence, pp. 52–60. Morgan Kaufmann (1991)
4. Castelo, R., Feelders, A., Siebes, A.: MAMBO: discovering association rules based on conditional independencies. In: Hoffmann, F., Hand, D.J., Adams, N., Fisher, D., Guimaraes, G. (eds.) IDA 2001. LNCS, vol. 2189, pp. 289–298. Springer, Heidelberg (2001). https://doi.org/10.1007/3-540-44816-0_29
5. Castelo, R., Guigó, R.: Splice site identification by idlBNs. Bioinformatics **20**(Suppl 1), i69–i76 (2004). https://doi.org/10.1093/bioinformatics/bth932
6. Castelo, R., Kočka, T.: On inclusion-driven learning of Bayesian networks. J. Mach. Learn. Res. **4**(Sep), 527–574 (2003)

7. Castelo, R., Siebes, A.: Priors on network structures. Biasing the search for Bayesian networks. Int. J. Approximate Reasoning **24**(1), 39–57 (2000). https://doi.org/10.1016/S0888-613X(99)00041-9

8. Castelo, R., Siebes, A.: A characterization of moral transitive acyclic directed graph Markov models as labeled trees. J. Stat. Plann. Inference **115**(1), 235–259 (2003). https://doi.org/10.1016/S0378-3758(02)00143-X

9. Castelo, R., Wormald, N.: Enumeration of P_4-free chordal graphs. Graphs Combin. **19**(4), 467–474 (2003). https://doi.org/10.1007/s00373-002-0513-9

10. Chickering, D.M.: A transformational characterization of equivalent Bayesian networks. In: Besnard, P., Hanks, S. (eds.) Proceedings of the Conference on Uncertainty in Artificial Intelligence, pp. 87–98. Morgan Kaufmann (1995)

11. Chickering, D.M.: Optimal structure identification with greedy search. J. Mach. Learn. Res. **3**(Nov), 507–554 (2002)

12. Chickering, D.M., Meek, C.: Finding optimal Bayesian networks. In: Proceedings of the Conference on Uncertainty in Artificial Intelligence, pp. 94–102 (2002)

13. Cooper, G., Herskovits, E.: A Bayesian method for the induction of probabilistic networks from data. Mach. Learn. **9**, 309–405 (1992)

14. Cooper, G.F., Yoo, C.: Causal discovery from a mixture of experimental and observational data. In: Proceedings of the Conference on Uncertainty in Artificial Intelligence, pp. 116–125 (1999)

15. Cox, D.R., Wermuth, N.: Multivariate Dependencies: Models, Analysis and Interpretation. Chapman and Hall/CRC (1996)

16. Friedman, N., Nachman, I., Peér, D.: Learning Bayesian network structure from massive datasets: the «sparse candidate» algorithm. In: Proceedings of the Conference on Uncertainty in Artificial Intelligence, pp. 206–215 (1999)

17. Giudici, P., Castelo, R.: Association models for web mining. Data Min. Knowl. Disc. **5**, 183–196 (2001)

18. Giudici, P., Castelo, R.: Improving Markov chain Monte Carlo model search for data mining. Mach. Learn. **50**, 127–158 (2003)

19. Harary, F.: Graph Theory. Addison-Wesley, London (1969)

20. Hauser, A., Bühlmann, P.: Characterization and greedy learning of interventional Markov equivalence classes of directed acyclic graphs. J. Mach. Learn. Res. **13**(1), 2409–2464 (2012)

21. Hauser, A., Bühlmann, P.: Jointly interventional and observational data: estimation of interventional Markov equivalence classes of directed acyclic graphs. J. R. Stat. Soc. Ser. B Stat Methodol. **77**(1), 291–318 (2015)

22. Heckerman, D., Geiger, D., Chickering, D.M.: Learning Bayesian networks: the combination of knowledge and statistical data. Mach. Learn. **20**, 197–243 (1995)

23. Kalisch, M., Bühlman, P.: Estimating high-dimensional directed acyclic graphs with the PC-algorithm. J. Mach. Learn. Res. **8**(3) (2007)

24. Kalisch, M., Mächler, M., Colombo, D., Maathuis, M.H., Bühlmann, P.: Causal inference using graphical models with the R package pcalg. J. Stat. Softw. **47**, 1–26 (2012)

25. Kočka, T., Castelo, R.: Improved learning of Bayesian networks. In: Breese, J., Koller, D. (eds.) Proceedings of the Conference on Uncertainty in Artificial Intelligence, pp. 269–276. Morgan Kaufmann (2001)

26. Kočka, T., Bouckaert, R., Studený, M.: On characterizing inclusion of Bayesian networks. In: Breese, J., Koller, D. (eds.) Proceedings of the Conference on Uncertainty in Artificial Intelligence, pp. 261–268. Morgan Kaufmann (2001)

27. Kuipers, J., Moffa, G., Heckerman, D.: Addendum on the scoring of Gaussian directed acyclic graphical models. Ann. Stat. **42**(4), 1689–1691 (2014)

28. Lauritzen, S.L.: Graphical Models. Oxford University Press, Oxford (1996)
29. Madigan, D., York, J.: Bayesian graphical models for discrete data. Int. Stat. Rev. 215–232 (1995)
30. Meek, C.: Graphical models, selecting causal and statistical models. Ph.D. thesis, Carnegie Mellon University (1997)
31. Mukherjee, S., Speed, T.P.: Network inference using informative priors. Proc. Natl. Acad. Sci. **105**(38), 14313–14318 (2008)
32. Nielsen, J.D., Kočka, T., Peña, J.M.: On local optima in learning Bayesian networks. In: Kjærulff, U., Meek, C. (eds.) Proceedings of the Conference on Uncertainty in Artificial Intelligence, pp. 435–442. Morgan Kaufmann (2003)
33. Pearl, J.: Probabilistic Reasoning in Intelligent Systems. Morgan Kaufmann, San Mateo (1988)
34. Pearl, J.: Causal diagrams for empirical research. Biometrika **82**(4), 669–688 (1995)
35. Pearl, J.: Causality: Models, Reasoning and Inference. Cambridge University Press (2009)
36. Robinson, R.W.: Counting labeled acyclic digraphs. In: Harary, F. (ed.) New Directions in the Theory of Graphs, pp. 239–273. Academic Press, New York (1973)
37. Schwarz, G.: Estimating the dimension of a model. Ann. Stat. 461–464 (1978)
38. Silander, T., Myllymäki, P.: A simple approach for finding the globally optimal Bayesian network structure. In: Dechter, R., Richardson, T. (eds.) Proceedings of the Conference on Uncertainty in Artificial Intelligence, pp. 445–452. Morgan Kaufmann (2006)
39. Spirtes, P., Glymour, C.: An algorithm for fast recovery of sparse causal graphs. Soc. Sci. Comput. Rev. **9**(1), 62–72 (1991)
40. Spirtes, P., Glymour, C., Scheines, R.: Causation, Prediction and Search. Springer, New York (1993)
41. Tsamardinos, I., Brown, L.E., Aliferis, C.F.: The max-min hill-climbing Bayesian network structure learning algorithm. Mach. Learn. **65**, 31–78 (2006)
42. Verma, T., Pearl, J.: Influence diagrams and d-separation. Technical report CSD 880052, R-101, Cognitive Systems Laboratory, UCLA (1988)
43. Verma, T., Pearl, J.: Equivalence and synthesis of causal models. In: Bonissone, P., Henrion, M., Kanal, L., Lemmer, J. (eds.) Proceedings of the Conference on Uncertainty in Artificial Intelligence, pp. 255–268. Morgan Kaufmann (1990)
44. Whittaker, J.: Graphical Models in Applied Multivariate Statistics. Wiley, New York (1990)
45. Wright, S.: Correlation and causation. J. Agric. Res. **20**(7), 557–585 (1921)

Large Language Models

From Symbolic to Neural and Back: Exploring Knowledge Graph–Large Language Model Synergies

Blaž Škrlj[1]([✉])(iD), Boshko Koloski[1,2](iD), Senja Pollak[1](iD), and Nada Lavrač[1](iD)

[1] Jožef Stefan Institute, Jamova cesta 39, 1000 Ljubljana, Slovenia
{blaz.skrlj,boshko.koloski,senja.pollak,nada.lavrac}@ijs.si
[2] Jožef Stefan International Postgraduate School, Jamova cesta 39,
1000 Ljubljana, Slovenia

Abstract. Integrating structured knowledge from Knowledge Graphs (KGs) into Large Language Models (LLMs) enhances factual grounding and reasoning capabilities. This survey paper systematically examines the synergy between KGs and LLMs, categorizing existing approaches into two main groups: KG-enhanced LLMs, which improve reasoning, reduce hallucinations, and enable complex question answering; and LLM-augmented KGs, which facilitate KG construction, completion, and querying. Through comprehensive analysis, we identify critical gaps and highlight the mutual benefits of structured knowledge integration. Compared to existing surveys, our study uniquely emphasizes scalability, computational efficiency, and data quality. Finally, we propose future research directions, including neuro-symbolic integration, dynamic KG updating, data reliability, and ethical considerations, paving the way for intelligent systems capable of managing more complex real-world knowledge tasks.

Keywords: Knowledge Graphs · Large Language Models · Natural Language Processing · Knowledge Discovery

1 Introduction

Knowledge Graphs and Large Language Models represent two foundational *Artificial Intelligence* (AI) and *Natural Language Processing* (NLP) paradigms. On the one hand, **Knowledge Graphs** (KGs) provide a structured and semantically rich representation of information by encoding entities as nodes and their interactions as edges. This graph-based representation enables computational systems to traverse complex information networks, facilitating improved data interpretation and contextualized reasoning. Such capabilities make KGs indispensable in domains that require a deep understanding of interconnected data, including semantic search, recommendation systems, drug design and comprehensive knowledge management [23,43,62]. On the other hand, **Large Language Models** (LLMs), exemplified by systems such as GPT-3 [8], employ

© The Author(s), under exclusive license to Springer Nature Switzerland AG 2026
M. van Leeuwen and J. Vreeken (Eds.): Arno Siebes Festschrift, LNCS 16067, pp. 181–197, 2026.
https://doi.org/10.1007/978-3-032-03028-3_11

advanced deep learning architectures to perform sophisticated linguistic tasks. These models are superior in natural language generation, comprehension and translation, relying on large-scale corpora and state-of-the-art neural architectures, most notably **Transformer networks** [57]. Through the assimilation of large amounts of data, LLMs are capable of mastering syntax, semantics, and contextual dependencies, achieving unprecedented performance in tasks ranging from text summarization to conversational interfaces [66].

This work surveys the intersection of these two paradigms, structured Knowledge Graphs and foundational Language Models, and examines their complementary potentials. By integrating the structured and semantically coherent framework of KGs with the generative capabilities of LLMs, it becomes feasible to construct hybrid systems that are both contextually aware and computationally robust. On the one hand, the structured information encapsulated in KGs may serve as a guiding framework for LLMs, ensuring that generated language is accurate and semantically grounded. On the other hand, LLMs can help in the generation, augmentation, and maintenance of KGs by extracting pertinent information from unstructured textual sources and systematically organizing it into coherent graph structures. The integration of these technologies can advance numerous applications: enhanced natural language understanding, improved techniques for information retrieval [19], and sophisticated knowledge discovery methodologies [3] are just a few of the many promising use cases. The logical structure of KGs, when combined with the contextual reasoning of LLMs, paves the way for AI systems to achieve greater precision, richer semantic interactions, and advanced analytical capabilities. In the following sections, summarized in Fig. 1, this work systematically explores the construction and refinement of knowledge graphs, the underlying LLM architectures and the methodologies by which these two paradigms can be effectively combined, resulting in a conceptual framework and practical guiding principles for scholars and practitioners interested in exploiting the potential of KG and LLM integration. By undertaking this research, we seek to advance the field's understanding of these two individual AI paradigms, their cross-fertilization, and their capacity to jointly expand the boundaries of AI research and applications.

2   Knowledge Graphs

Mining knowledge graphs involves extracting, processing, and organizing data from various sources to build a comprehensive and structured graph representation. Key techniques include **entity recognition**, **entity linking**, **relationship extraction**, and **data integration** [22,67].

Entity recognition involves identifying and classifying entities (e.g., people, organizations, locations) within text. This step is crucial for populating knowledge graphs with accurate and relevant data. Advanced techniques, such as named entity recognition (NER) models, leverage machine learning and natural language processing to achieve high accuracy in identifying entities [22,51]. These models can be trained on large annotated datasets to recognize various types of entities in different domains. As an illustration, Table 1 presents

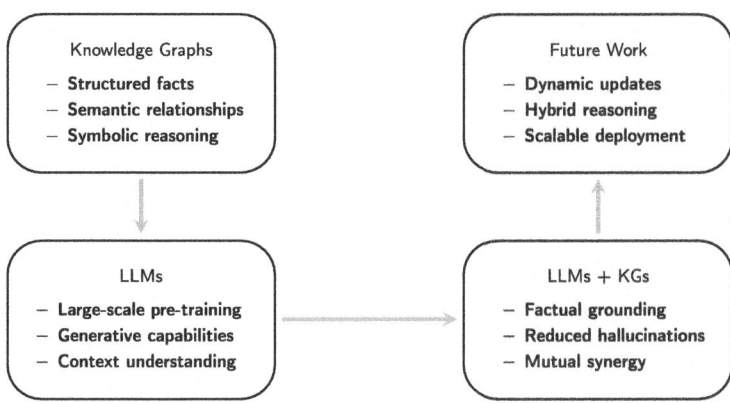

Fig. 1. Scope of this paper, illustrating the evolution from Knowledge Graphs to LLMs, their synergy, and directions for future work.

a random sample of triplets related to the concept "mountain", taken from the ConceptNet [53] knowledge graph.

Relationship extraction focuses on identifying and categorizing the relationships between entities, such as "works at" or "located in". This process is essential for building the connections within the graph that represent the interactions and associations between entities. Techniques for relationship extraction include rule-based methods, supervised learning algorithms, and deep learning models [18]. These methods can analyze the context and syntactic structure of sentences to accurately determine the nature of relationships between entities.

Entity linking focuses on extracting entities from free-form text and correctly mapping them to the corresponding entries in a knowledge graph, such as Wikidata. For example, in the sentence "Jeff Bezos founded Amazon", entity linking maps "Amazon" to the company Amazon (Wikidata entity Q3884) and "Jeff Bezos" to the person Jeff Bezos (Wikidata entity Q312556). Entity linking methods vary from specialist models to generalist models that perform simulta-

Table 1. A random sample of head-relation-tail triplets related to the concept "mountain" from the ConceptNet knowledge graph [53].

head	relation	tail
`/c/fr/eucalyptus_camphora/n/wn/plant`	`/r/Synonym`	`/c/en/mountain_swamp_gum/n/wn/plant/`
`/c/en/mountain_bike/n`	`/r/RelatedTo`	`/c/en/traction/`
`/c/en/faith_will_move_mountains`	`/r/RelatedTo`	`/c/en/god/`
`/c/es/diablo_espinoso/n/wn/animal`	`/r/Synonym`	`/c/en/mountain_devil/n/wn/animal/`
`/c/en/antiapennine/n`	`/r/RelatedTo`	`/c/en/mountain_range/`
`/c/en/climbing_mountain`	`/r/UsedFor`	`/c/en/reaching_top/`
`/c/pl/robić_z_igły_widły/v`	`/r/RelatedTo`	`/c/en/make_mountain_out_of_molehill/`

neous relation extraction and entity linking, with state-of-the-art models being built on transformer architecture [52].

Data integration combines information from multiple sources to create a unified knowledge graph. This step involves aligning and merging data from diverse datasets, resolving conflicts, and ensuring consistency. Data integration techniques include schema matching, entity resolution, and data fusion [22]. Schema matching aligns different data schemas to a common representation, while entity resolution identifies and merges duplicate entities across datasets. Data fusion combines information from multiple sources, selecting the most reliable data, and resolving inconsistencies.

Knowledge graphs are used in a wide range of applications, including semantic search, recommendation systems, and question-answering systems. Semantic search improves the accuracy of search engines by understanding the meaning behind queries. Unlike traditional keyword-based search, semantic search leverages relationships and context within knowledge graphs to deliver more relevant results [47]. For example, a query about "famous scientists in the 20th century" can return results that include notable figures and their contributions, even if the exact keywords are not present in the search query. **Recommendation systems** use knowledge graphs to provide personalized suggestions based on user preferences and behaviors. By analyzing the relationships between users, items, and various attributes, knowledge graphs enable recommendation systems to offer more accurate and relevant recommendations [21]. For instance, a movie recommendation system can suggest films based on a user's viewing history, preferences, and the relationships between movies, actors, and genres within the knowledge graph. Question-answering systems leverage knowledge graphs to deliver precise and contextually relevant responses to user queries. These systems can traverse the graph to find the most accurate information, considering the relationships and connections between entities [17]. For example, a question about "the capital of France" can be answered by navigating the knowledge graph to find the entity "France" and its associated attribute "capital".

To address the rigidity and sparsity of purely symbolic knowledge graphs, we turn to knowledge graph embeddings (KGEs), which map entities and relations into continuous vector spaces and enable efficient numerical reasoning for downstream machine learning tasks such as link prediction, entity classification, and clustering. Popularised by the representational learning paradigm [6], KGEs facilitate the application of symbolic knowledge to tasks such as text classification [28] and enrich representations for tabular data [27], enabling efficient interaction between structured semantic knowledge and diverse downstream modalities.

Early models such as TransE and its variants (TransH, TransR, TransD) treated relationships as translations in embedding space. More recent techniques, such as ConvE and RotatE, have improved on these by leveraging convolutional networks and rotational transformations to capture complex relational patterns [20,59]. Training strategies significantly impact KGE effectiveness. Ruffinelli et al. [50] demonstrated that with proper hyperparameter tuning, even older models like RESCAL and TransE can match or outperform newer architectures. A major direction in embedding research is the incorporation of relation paths, as shown in PConvKB, which integrates path-based attention to improve embeddings [26]. Another breakthrough, PairRE, introduced the concept of paired relation vectors to enhance the representation of N-to-N and 1-to-N relationships [9].

Recent advances continue to refine KGE methods. Li and Zhu [31] proposed TransE-MTP, which builds on TransE but incorporates **multi-translation principles** to better capture hierarchical and compositional structures. The study of **benchmarking KGE models** has also gained traction, with systematic evaluations highlighting efficiency vs. effectiveness trade-offs across models [20]. Mining knowledge graphs involves a combination of **entity recognition**, **relationship extraction**, and **data integration** techniques to build a structured representation of information. These knowledge graphs are invaluable in various applications, enhancing the accuracy and relevance of search, recommendation, and question-answering systems.

However, KGs present several inherent challenges. Often building and maintaining their structure demands domain experts—a costly and often scarce resource—especially when handling temporal information, which complicates entity linking and versioning. Additionally, most efforts rely on a fixed ontology, making later revisions labor-intensive and error-prone. Real-world data scarcity and heterogeneity further hinder graph construction, as missing or inconsistent data require careful mapping and manual correction. Finally, merging graphs or expanding them across diverse subdomains frequently exposes mismatches in terminology, granularity, and temporal scope, undermining seamless integration.

3   Large Language Models (LLMs)

Large Language Models (LLMs) represent a significant breakthrough in natural language processing (NLP) and artificial intelligence (AI), enabling machines to understand, generate, and interact using human language with remarkable fluency. These models leverage deep learning architectures to process large amounts of textual data, learning intricate linguistic patterns, factual knowledge, and contextual relationships. One of the most notable advances in this domain has been the adoption of the **Transformer architecture** [57], which introduced self-attention mechanisms that allow models to capture long-range dependencies efficiently.

Modern LLMs, such as OpenAI's GPT-3, which boasts 175 billion parameters [8], have demonstrated unprecedented capabilities across a wide range of language-related tasks, including text completion, machine translation, summarization, question-answering, and even creative writing. These models undergo a two-step process: pre-training on a large corpus of diverse text followed by fine-tuning on specific tasks. The pre-training phase is often conducted using self-supervised learning objectives such as masked language modeling (MLM) in BERT [16] or autoregressive token prediction in GPT [48]. The fine-tuning phase further refines the model using supervised datasets tailored to domain-specific tasks, significantly improving performance. LLMs are referred to as **foundation models** [7], which means that they serve as a general-purpose base upon which various downstream applications can be built. The scalability of these models, following well-documented scaling laws [8], suggests that increasing model size, dataset size, and compute resources leads to predictable improvements in language understanding and generation. However, these models also raise challenges concerning ethical AI, bias mitigation, interpretability, and computational efficiency, topics that continue to be actively researched.

To understand how LLMs achieve their capabilities, we explore their fundamental components, including attention mechanisms, token embeddings, and positional encodings. Attention mechanisms have been a transformative innovation in NLP, allowing models to dynamically weigh the importance of different words in a given input sequence [4]. This idea was first introduced in the context of neural machine translation, where it enabled models to attend to relevant words in the source sentence while generating the target sentence. The Transformer architecture expanded upon this idea with the introduction of *self-attention* [57], where each word attends to every other word in the sequence, regardless of distance. The Transformer's multi-head self-attention mechanism further enhances this capability by computing multiple attention scores in parallel, each capturing different linguistic relationships [57]. This is a fundamental departure from recurrent architectures, which struggle with long-range dependencies due to vanishing gradient issues. The ability to capture complex syntactic and semantic relationships has made the Transformer the de facto standard for modern NLP models.

Words and phrases in human language need to be represented in a numerical form that models can process. This is achieved through token embeddings, which map words, subwords, or characters into dense vector spaces. Early approaches to word embeddings, such as Word2Vec [37] and GloVe [44], demonstrated that words with similar meanings often have similar vector representations. While early word embeddings were static (i.e. the representation of a word was fixed regardless of context), modern LLMs leverage **contextual embeddings**, where the meaning of a word adapts based on surrounding words. This advancement, introduced by models like BERT [16], allows disambiguation of homonyms and polysemous words. For example, in the sentence "The bank is located near the river", the meaning of "bank" is different from its meaning in "I deposited money in the bank". Contextual embeddings ensure that each instance of "bank" receives a different vector representation depending on its usage. LLMs have revolutionized numerous NLP tasks, spanning both general and domain-specific applications. Some of the most significant use cases are listed below. The Transformer architecture has significantly advanced machine translation, outperforming earlier sequence-to-sequence recurrent models [57]. By leveraging self-attention, LLMs capture **long-range dependencies** and subtle linguistic nuances, enabling accurate translation across multiple languages. Google's Transformer-based translation models have set new benchmarks in translation quality, achieving human-like fluency in certain language pairs. Text

Table 2. Selected top 10 Open Source LLMs with Parameter Counts and Licenses.

Model	License	Developer	Parameter Count	Year	BibTeX Citation
LLaMA 3.1	Community	Meta AI	405B	2024	[1]
Mixtral 8x22B	Apache 2.0	Mistral AI	141B	2024	[2]
Command R	Open Source (Research)	Cohere	35B	2024	[12]
Command R+	Open Source (Research)	Cohere	103B	2024	[13]
Gemma	Apache 2.0	Google DeepMind	7B	2024	[14]
Jamba	Apache 2.0	AI21 Labs	52B	2024	[29]
DBRX	Apache 2.0	Databricks Mosaic ML	132B	2024	[38]
Phi-3 Medium	MIT	Microsoft	14B	2024	[36]
Nemotron-4	Apache 2.0	Nvidia	340B	2024	[39]
DeepSeek-V3	MIT	DeepSeek	671B	2024	[15]

summarization is essential for condensing lengthy documents into concise versions while preserving key information. Models like BERTSUM [16] and T5 [49] have demonstrated remarkable proficiency in extractive and abstractive summarization. Extractive summarization selects the most relevant sentences from a document, while abstractive summarization generates new text that captures the original meaning concisely. Sentiment analysis involves determining the emotional tone of text, which is widely used in social media monitoring, customer feedback analysis, and political discourse analysis. Transformer-based models fine-tuned on sentiment classification datasets, such as IMDb movie reviews or Twitter sentiment datasets, achieve state-of-the-art accuracy [16].

One of the most groundbreaking applications of LLMs is in conversational AI, where models like GPT-3 [8] and Google's LaMDA [55] generate human-like dialogue. These models are used in chatbots, virtual assistants, and interactive agents, enhancing customer service, education, and entertainment experiences. LLMs have also been tailored for domain-specific applications, such as scientific literature analysis and biomedical NLP. SciBERT [5], trained on scientific papers, excels at processing technical language, while BioBERT, specialized for biomedical texts, enhances tasks like named entity recognition and question answering in medicine. Recently, open source movement targeting accessible language models has given rise to many high-quality models. Selected examples are shown in Table 2.

4   KG-LLM Integration

The intersection of LLMs and KGs has emerged as a promising area of research, aimed at combining the generalization power of LLMs with the precision and symbolic reasoning of KGs. The integration may strengthen the strengths, and weaken the weaknesses of the individual technologies:

- LLMs are at the forefront of AI-driven language understanding, shaping the future of human-computer interaction and knowledge discovery. Large Language Models (LLMs) such as GPT-4 have achieved impressive results in

Table 3. Coverage of topics in recent LLM-and-KG surveys: Pan et al. (2024) [42], Pan et al. (2023) [40], Hu et al. (2023) [24], Yang et al. (2024) [61]. ✓ = topic is covered; × = not covered; − = out of scope/not applicable.

Topic	Pan et al. (2024)	Pan et al. (2023)	Hu et al. (2023)	Yang et al. (2024)
Entity Linking & Alignment	✓	✓	−	×
Relation & Attribute Extraction	✓	✓	−	✓
KG Embedding (Representation)	✓	✓	×	×
KG Completion (Link Prediction)	✓	✓	×	✓
Graph-to-Text Generation	✓	×	×	×
KG Question Answering	✓	✓	×	✓
KG-Enhanced LLMs (Knowledge-Injected)	✓	✓	✓	✓
LLM-Augmented KGs (KG tasks)	✓	✓	−	−
Synergized LLM+KG (Bidirectional)	✓	✓	×	✓
Hallucination & Factual Accuracy	✓	✓	×	✓

generating human-like text and answering questions across domains. However, purely neural LLMs often act as black-boxes and may **hallucinate** incorrect facts or lack access to up-to-date knowledge.
- In contrast, a Knowledge Graph (KG) is a structured repository of facts represented as a graph of entities and relationships; KGs offer interpretability and reliable information, but are typically incomplete and costly to curate.

By integrating LLM and KG technologies, we can build systems that answer questions and generate content with both the fluency of natural language and the correctness of evidence-backed knowledge. Multiple recent works have explored different aspects of KG-LLM integration. For example, Pan et al. [42] present a roadmap for unifying LLMs and KGs, describing frameworks where KGs assist LLMs and vice versa. Another survey by Pan et al. [40] discusses key opportunities and challenges when bridging these models. Hu et al. [24] focus on techniques for injecting knowledge (often from KGs) into pre-trained language models, and Yang et al. [61] propose methods to enhance LLM factual consistency using KG-based information. Each of these studies covers a subset of the full landscape. Table 3 provides a comparative overview of the topics addressed by these surveys, pointing the interested reader to the topics of relevance. We consider as non-applicable the topics which either weren't yet relevant at the time of writing a paper, or intentionally focus on a different topic (it's apparent there is no coverage).

Building on the above paradigms, we organize the overview of the paradigms at the intersection of KGs and LLMs into three primary categories, summarized in Fig. 2. First, **KG-enhanced LLMs** use knowledge graphs to improve the performance of language models. Second, **LLM-augmented KGs** leverage language models to construct, enrich, or utilize knowledge graphs. Third, **joint LLM–KG frameworks** involve a deeper bidirectional coupling of LLMs and KGs. We next describe each of these categories, along with representative use cases and techniques.

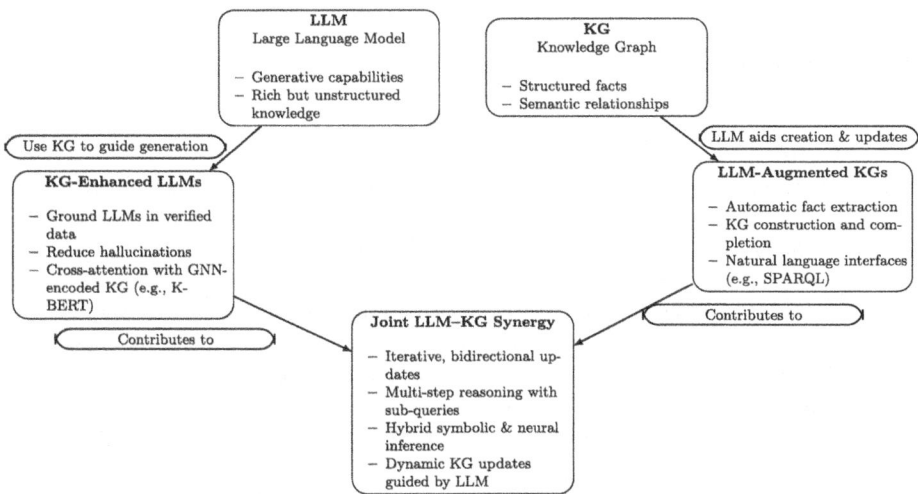

Fig. 2. The interplay between Large Language Models (LLMs) and Knowledge Graphs (KGs).

4.1   KG-Enhanced LLMs

KG-enhanced LLMs represent an emerging research paradigm that seeks to bridge the gap between vast but sometimes imprecise knowledge stored in large language models (LLMs) and the structured curated information contained in knowledge graphs (KGs). In this approach, a KG acts as an external reliable repository of factual data that the LLM can query to enrich its responses or guide its reasoning process. For example, by retrieving relevant subgraphs from a KG such as Wikidata, DBpedia, or ConceptNet, an LLM-based question-answering system can incorporate precise facts (such as the familial relationships of a historical figure) into its generated output, to enhance factual consistency and reduce hallucinations [61]. Several integration strategies have been explored. A common method is to augment the input prompt with additional context extracted from a KG, to effectively ground the LLM's generation in verified data. Alternatively, more sophisticated techniques embed KG-derived representations directly into the model's latent space. In these approaches, adapter modules or fine-tuning procedures are used to inject structured knowledge into the LLM's parameters, helping the model internalize and recall factual information more reliably [24].

Recent work has also explored the use of graph neural networks (GNNs) to encode retrieved KG subgraphs, which are then integrated with LLMs via cross-attention mechanisms [63]. Such techniques enable a form of "knowledge-guided reasoning", where intermediate model representations are directly influenced by structured data from the KG. Early models like K-BERT [33] and KnowBERT [45] (originally developed to enhance language representations) have inspired further innovations in this direction. Moreover, models such as ERNIE [54] have demonstrated that pre-training on knowledge-enhanced corpora can yield significant improvements in downstream tasks, particularly in question-answering and logical inference. Beyond language-only models, [11] recently demonstrated that KGs can be introduced to Vision-Language Models (VLMs) as a fallback mechanism when a VLM is uncertain and can be used as a knowledge base to further improve visual question answering.

KG-enhanced LLMs combine the generative versatility of LLMs with the precision of structured knowledge sources. This fusion not only mitigates issues like hallucinations but also improves the interpretability of outputs by offering traceable reasoning paths through the underlying KG. Another recent attempt that exploits KG-based prompting to improve language models that operate via chain of thought (CoT) is the MindMap project [60]. This project showcases structured KG prompting techniques that improve LLM reasoning accuracy and reduce hallucinations. As research in this area continues to mature, we expect further innovations that refine the synergy between unstructured language models and structured knowledge graphs.

4.2   LLM-Augmented KGs

In the opposite direction, researchers have employed LLMs to assist in creating, populating, and querying knowledge graphs. LLMs can serve as powerful information extractors: given large corpora of text, an LLM can be prompted or fine-tuned to identify entities and relationships, essentially converting unstructured text into structured triples for **KG construction** [10, 46]. This approach has been used to rapidly expand KGs by mining textual resources for new facts, significantly reducing the manual effort typically needed for KG curation. However, one needs to be considered when using LLMs to populate KGs, as LLM-augmented KGs run the risk of propagating the model's hallucinations and biases into the KG. Thus, rigorous validation and human-in-the-loop curation are essential to ensure the integrity and reliability of these automatically generated knowledge graphs. Furthermore, **KG completion** tasks (predicting missing links or attributes in an existing graph), LLMs can leverage their broad world knowledge to suggest likely connections between entities [42]. Another application is using LLMs to translate natural language queries or commands into formal queries that a KG can execute (e.g., converting a question into SPARQL query language). By acting as an intelligent interface, the LLM makes the KG accessible to users who do not know the KG's query language, enabling conversational **question answering** over the graph. Additionally, LLMs have been utilized for **KG-to-text generation** to produce coherent descriptions or summaries of specific entities or subgraphs. Such generated text can help end users understand KG data in natural language and can also serve as supplementary training data for language models. Recent research has further extended the capabilities of LLM-augmented KGs by integrating explicit structural cues into the language models. For example, Zhang et al. [64] proposed a novel method where structural embeddings of entities and relations are injected into the LLM via a knowledge prefix adapter. This technique, named **structure-aware reasoning**, enables the LLM to combine textual context with the inherent graph structure, enhancing its ability to more accurately predict missing links in knowledge graph completion. By merging cross-modal structural information with raw text, the approach effectively mitigates issues such as hallucinations and improves the reliability of KG completion. Another study by Zhu et al. [68] offers a comprehensive evaluation of LLMs for KG construction and reasoning. Their work demonstrates that while models like GPT-4 excel in general language understanding, their performance in structured tasks such as link prediction and multi-hop query answering can be significantly boosted through domain-specific prompt engineering and targeted fine-tuning. The study shows that LLMs can not only extract entities and relations but also perform complex reasoning over these structures, thereby supporting applications that require in-depth inference and dynamic knowledge retrieval.

The overviewed work indicates that LLMs can be used as external domain evaluators, inspecting the quality of knowledge graphs extracted from raw, unstructured text, performing on par with or superseding human performance [25,56]. These advances illustrate that the synergy between LLMs and knowledge graphs is evolving beyond simple extraction and query translation. By incorporating structural embeddings and reasoning strategies, LLM-augmented KGs are becoming more robust and interactive.

5   Joint LLM–KG Synergy

Large Language Models (LLMs) and Knowledge Graphs (KGs) have complementary strengths, motivating frameworks that tightly integrate the two for bidirectional reasoning [41]. Recent works propose joint architectures where an LLM's neural reasoning is coupled with a KG's symbolic structure in multi-step, interactive loops. For example, Liu et al. present a dual reasoning paradigm pairing an LLM with a graph neural network (GNN) to perform collaborative multi-hop reasoning [32]. In their framework, the GNN derives explicit relational chains from the KG, providing interpretable paths subsequently converted into a knowledge-enhanced prompt guiding the LLM's inference [32]. This division of labor—with symbolic traversal handled by the KG and nuanced language understanding by the LLM—achieves state-of-the-art results in knowledge-intensive question answering (QA), demonstrating how mutual reinforcement can curb hallucinations and improve correctness.

Other approaches grant LLMs direct capabilities to query and update KGs **during reasoning**. Markowitz et al. introduce Tree-of-Traversals, a zero-shot *neuro-symbolic planning* algorithm equipping a black-box LLM with discrete actions to interface with a KG [35]. At each step, the LLM decides either to *expand a node* via a relation in the KG or to *propose an answer*, dynamically constructing a reasoning tree. A search procedure evaluates and revisits these actions, ensuring high-confidence reasoning paths [35]. This tight loop allows KG information to influence the LLM's reasoning dynamically and vice versa, significantly improving multi-hop QA performance without fine-tuning. Similarly, Li et al. propose Decoding-on-Graphs (DoG), constraining the LLM's reasoning chain to conform to actual KG paths [30]. Applying graph-derived constraints during decoding ensures that each reasoning step is grounded in factual KG context, yielding faithful and sound multi-step reasoning that is both interpretable and effective for KG-based QA.

Another research direction employs KGs as navigation guides for LLM retrieval and reasoning. Ma et al. introduce Think-on-Graph 2.0, a retrieval-augmented generation framework aligning queries with KG structure to direct information retrieval [34]. Here, the KG functions as a map, helping the LLM select entities and relations for detailed textual evidence retrieval. This KG-guided retrieval enhances logical consistency and reduces errors in multi-hop queries, highlighting the efficacy of hybrid structured–neural systems. Additionally, KG-derived planning has been harnessed to enhance LLM step-by-step reasoning. Wang et al. devise a method to train LLMs with planning data extracted from KGs, improving their iterative retrieval and reasoning capabilities through targeted fine-tuning [58]. Emerging studies also explore mutual refinement, allowing continuous reciprocal updates between LLMs and KGs. Zhang et al. introduce Chain-of-Knowledge, integrating symbolic inference into LLMs by training on logical rules mined from KGs [65]. This iterative symbolic–neural co-training strengthens logical consistency within LLM reasoning processes, yielding superior performance on reasoning benchmarks.

Cutting-edge research is moving towards tight integration of LLMs and KGs, from pipeline architectures injecting KG-derived facts into LLM prompts, interactive agents interweaving language generation and graph traversal, to joint training techniques aligning neural representations with symbolic logic. This joint synergy supports dynamic bidirectional interaction—KGs grounding LLM reasoning steps, and LLMs populating or navigating KGs—resulting in robust, explainable, and knowledgeable AI systems [35,41].

6   Interesting Open Problems for Future Research

Based on the reviewed work, we have identified a list of open problems in the field of combining LLMs and KGs.

1. **Structured-Unstructured Alignment:** How can we seamlessly align structured information from knowledge graphs with the unstructured, free-form text generated by LLMs to enable robust joint reasoning?
2. **Neuro-Symbolic Integration:** How can we optimally fuse neural representations with symbolic reasoning, combining the strengths of both paradigms for more effective problem solving?
3. **Dynamic Knowledge Updating:** How can joint LLM–KG frameworks support continuous, automated updates of the knowledge graph, incorporating new information and corrections over time?
4. **Data Quality and Consistency:** What techniques can ensure that automatically extracted or augmented knowledge maintains high accuracy and consistency across diverse and evolving data sources?

5. **Hallucination Mitigation:** What novel approaches can leverage KG constraints to reduce LLM hallucinations and improve factual grounding in generated content?
6. **Scalability and Efficiency:** How can we design integrated systems that effectively scale both large language models and massive knowledge graphs, while managing computational and memory overhead?
7. **Interpretability and Explainability:** What methods can improve the interpretability of hybrid systems so that reasoning paths, which span both symbolic (KG) and neural (LLM) components, become transparent and traceable?

7   Final Remarks

We explore the integration of knowledge graphs (KGs) and large language models (LLMs), reviewing core methods such as relation extraction in KGs and self-attention architectures in LLMs. Our analysis highlights that combining structured KGs with context-aware LLMs enhances factual accuracy and consistency in language generation, while LLM-augmented KGs automate knowledge graph construction and maintenance. A comparative analysis with existing surveys identifies coverage gaps and outlines key areas needing further investigation. Despite encouraging results, challenges persist, including scalability, computational efficiency, continuous data quality, and ethical concerns like model bias. Future research should aim to optimize KG–LLM integration, advance neurosymbolic methods, refine automated knowledge extraction, and establish standardized evaluation benchmarks.

Acknowledgments. We acknowledge the financial support of the Slovenian Research and Innovation Agency (ARIS) through grants GC-0002 (Large Language Models for Digital Humanities), GC-0001 (Artificial Intelligence for Science) and the core research programme P2-0103 (Knowledge Technologies). The work of B.K. was supported by the Young Researcher Grant PR-12394. We thank the anonymous reviewers for the constructive feedback. This work was partially funded by ARIS project with reference number J4-4555.

References

1. Meta AI. Llama 3.1: a 405B-parameter open-source language model (2024). https://github.com/facebookresearch/llama. Community license
2. Mistral AI. Mixtral 8x22b: sparse mixture-of-experts model with 141B parameters (2024). https://github.com/mistralai/Mixtral. Apache 2.0 License
3. Alam, M.M., et al.: Language model guided knowledge graph embeddings. IEEE Access **10**, 76008–76020 (2022)

4. Bahdanau, D., Cho, K., Bengio, Y.: Neural machine translation by jointly learning to align and translate. In: Bengio, Y., LeCun, Y. (eds.) 3rd International Conference on Learning Representations, ICLR 2015, San Diego, CA, USA, 7–9 May 2015, Conference Track Proceedings (2015)

5. Beltagy, I., Lo, K., Cohan, A.: SciBERT: a pretrained language model for scientific text. In: Inui, K., Jiang, J., Ng, V., Wan, X. (eds.) Proceedings of the 2019 Conference on Empirical Methods in Natural Language Processing and the 9th International Joint Conference on Natural Language Processing (EMNLP-IJCNLP), Hong Kong, China, pp. 3615–3620. Association for Computational Linguistics (2019)

6. Bengio, Y., Courville, A.C., Vincent, P.: Representation learning: a review and new perspectives. IEEE Trans. Pattern Anal. Mach. Intell. **35**(8), 1798–1828 (2013)

7. Bommasani, R., et al.: On the opportunities and risks of foundation models. ArXiv preprint, abs/2108.07258 (2021)

8. Brown, T.B., et al.: Language models are few-shot learners. In: Larochelle, H., Ranzato, M.A., Hadsell, R., Balcan, M.-F.,Lin, H.T. (eds.) Advances in Neural Information Processing Systems 33: Annual Conference on Neural Information Processing Systems 2020, NeurIPS 2020, 6–12 December 2020, Virtual (2020)

9. Chao, L., He, J., Wang, T., Chu, W.: PairRE: knowledge graph embeddings via paired relation vectors. InL Zong, C., Xia, F., Li, W., Navigli, R. (eds.) Proceedings of the 59th Annual Meeting of the Association for Computational Linguistics and the 11th International Joint Conference on Natural Language Processing (Volume 1: Long Papers), pp. 4360–4369. Association for Computational Linguistics (2021)

10. Chen, H., Shen, X., Lv, Q., Wang, J., Ni, X., Ye, J.: SAC-KG: exploiting large language models as skilled automatic constructors for domain knowledge graph. In: Ku, L.-W., Martins, A., Srikumar, V. (eds.) Proceedings of the 62nd Annual Meeting of the Association for Computational Linguistics (Volume 1: Long Papers), Bangkok, Thailand, pp. 4345–4360. Association for Computational Linguistics (2024)

11. Cocchi, F., Moratelli, N., Cornia, M., Baraldi, L., Cucchiara, R.: Augmenting multimodal LLMs with self-reflective tokens for knowledge-based visual question answering (2025)

12. Cohere. Command R: a retrieval-augmented open source language model (2024). https://github.com/cohere-ai/command-r. Open source for research

13. Cohere. Command R+: enhanced version for large-scale tasks (2024). https://github.com/cohere-ai/command-r-plus. Open source, extended capabilities

14. Google DeepMind. Gemma: Lightweight LLMs in 2B and 7B sizes (2024). https://github.com/google-research/gemma. Open source

15. DeepSeek. DeepSeek-v3: scaling open-source LLMs via efficient retrieval (2024). https://github.com/deepseek-ai/DeepSeek-V3. Open source

16. Devlin, J., Chang, M.-W., Lee, K., Toutanova, K.: BERT: pre-training of deep bidirectional transformers for language understanding. In: Burstein, J., Doran, C., Solorio, T. (eds.) Proceedings of the 2019 Conference of the North American Chapter of the Association for Computational Linguistics: Human Language Technologies, Volume 1 (Long and Short Papers), Minneapolis, Minnesota, pp. 4171–4186. Association for Computational Linguistics (2019)

17. Diefenbach, D., López, V., Singh, K., Maret, P.: Core techniques of question answering systems over knowledge bases: a survey. Knowl. Inf. Syst. **55**(3), 529–569 (2018)

18. Dong, X., et al.: Knowledge vault: a web-scale approach to probabilistic knowledge fusion. In: Macskassy, S.A., Perlich, C., Leskovec, J., Wang, W., Ghani, R. (eds.)

The 20th ACM SIGKDD International Conference on Knowledge Discovery and Data Mining, KDD 2014, New York, NY, USA, 24–27 August 2014, pp. 601–610. ACM (2014)

19. Fei, H., Ren, Y., Zhang, Y., Ji, D., Liang, X.: Enriching contextualized language model from knowledge graph for biomedical information extraction. Brief. Bioinform. **22**(3), bbaa110 (2021)

20. Ferrari, I., Frisoni, G., Italiani, P., Moro, G., Sartori, C.: Comprehensive analysis of knowledge graph embedding techniques benchmarked on link prediction. Electronics **11**(23), 3866 (2022)

21. Guo, Q., et al.: A survey on knowledge graph-based recommender systems. IEEE Trans. Knowl. Data Eng. **34**(8), 3549–3568 (2022)

22. Hogan, A., et al.: Knowledge Graphs. Synthesis Lectures on Data, Semantics, and Knowledge. Morgan & Claypool Publishers (2021)

23. Hogan, A., Blomqvist, E., Cochez, M., d'Amato, C., De Melo, G., Gutierrez, C., Kirrane, S., Gayo, J.E.L., Navigli, R., Neumaier, S., et al.: Knowledge graphs. ACM Comput. Surv. (Csur) **54**(4), 1–37 (2021)

24. Linmei, H., Liu, Z., Zhao, Z., Hou, L., Nie, L., Li, J.: A survey of knowledge-enhanced pre-trained language models. IEEE Trans. Knowl. Data Eng. **36**, 1413–1430 (2023)

25. Huang, H., Chen, C., He, C., Li, Y., Jiang, J., Zhang, W.: Can LLMs be good graph judger for knowledge graph construction? (2025)

26. Jia, N., Cheng, X., Su, S.: Improving knowledge graph embedding using locally and globally attentive relation paths. In: Jose, J.M., et al. (eds.) ECIR 2020. LNCS, vol. 12035, pp. 17–32. Springer, Cham (2020). https://doi.org/10.1007/978-3-030-45439-5_2

27. Kim, M.J., Grinsztajn, L., Varoquaux, G.: CARTE: pretraining and transfer for tabular learning. In: Proceedings of the 41st International Conference on Machine Learning, ICML 2024. JMLR.org (2024)

28. Koloski, B., Perdih, T.S., Robnik-Šikonja, M., Pollak, S., Škrlj, B.: Knowledge graph informed fake news classification via heterogeneous representation ensembles. Neurocomputing **496**, 208–226 (2022)

29. AI21 Labs. Jamba: A production-grade hybrid LLM (2024). https://github.com/ai21labs/Jamba. Open source under Apache 2.0

30. Li, K., Zhang, T., Wu, X., Luo, H., Glass, J., Meng, H.: Decoding on graphs: faithful and sound reasoning on knowledge graphs through generation of well-formed chains. arXiv preprint arXiv:2410.18415 (2024)

31. Li, Y., Zhu, C.: TransE-MTP: a new representation learning method for knowledge graph embedding with multi-translation principles and TransE. Electronics **13**(16), 3171 (2024)

32. Liu, G., Zhang, Y., Li, Y., Yao, Q.: Dual reasoning: a GNN-LLM collaborative framework for knowledge graph question answering. In: Conference on Parsimony and Learning (CPAL) (2025)

33. Liu, W., et al.: K-BERT: enabling language representation with knowledge graph. In: The Thirty-Fourth AAAI Conference on Artificial Intelligence, AAAI 2020, The Thirty-Second Innovative Applications of Artificial Intelligence Conference, IAAI 2020, The Tenth AAAI Symposium on Educational Advances in Artificial Intelligence, EAAI 2020, New York, NY, USA, 7–12 February 2020, pp. 2901–2908. AAAI Press (2020)

34. Ma, S., Xu, C., Jiang, X., Li, M., Qu, H., Guo, J.: Think-on-graph 2.0: deep and interpretable large language model reasoning with knowledge graph-guided retrieval. arXiv preprint arXiv:2407.10805 (2023)

35. Markowitz, E., et al.: Tree-of-traversals: a zero-shot reasoning algorithm for augmenting black-box language models with knowledge graphs. arXiv preprint arXiv:2407.21358 (2024)
36. Microsoft. Phi-3 medium: A 14B-parameter language model (2024). https://github.com/microsoft/phi. Open source under MIT License
37. Mikolov, T., Sutskever, I., Chen, K., Corrado, G.S., Dean, J.: Distributed representations of words and phrases and their compositionality. In: Burges, C.J.C., Bottou, L., Ghahramani, Z., Weinberger, K.Q. (eds.) Advances in Neural Information Processing Systems 26: 27th Annual Conference on Neural Information Processing Systems 2013. Proceedings of a Meeting Held December 5-8, 2013, Lake Tahoe, Nevada, United States, pp. 3111–3119 (2013)
38. Databricks Mosaic ML. DBRX: an open-source mixture-of-experts language model (2024). https://github.com/databricks/DBRX. Open source
39. Nvidia. Nemotron-4: A 340B-parameter open source language model (2024). https://github.com/nvidia/Nemotron-4. Apache 2.0 License
40. Pan, J.Z., et al.: Large language models and knowledge graphs: opportunities and challenges. Trans. Graph Data Knowl. **1**(1), 2:1–2:38 (2023)
41. Pan, S., Luo, L., Wang, Y., Chen, C., Wang, J., Wu, X.: Unifying large language models and knowledge graphs: a roadmap. arXiv preprint arXiv:2306.08302 (2023)
42. Pan, S., Luo, L., Wang, Y., Chen, C., Wang, J., Xindong, W.: Unifying large language models and knowledge graphs: a roadmap. IEEE Trans. Knowl. Data Eng. **36**(7), 3580–3599 (2024)
43. Peng, C., Xia, F., Naseriparsa, M., Osborne, F.: Knowledge graphs: opportunities and challenges. Artif. Intell. Rev. **56**(11), 13071–13102 (2023)
44. Pennington, J., Socher, R., Manning, C.: GloVe: global vectors for word representation. In: Moschitti, A., Pang, B., Daelemans, W. (eds.) Proceedings of the 2014 Conference on Empirical Methods in Natural Language Processing (EMNLP), Doha, Qatar, pp. 1532–1543. Association for Computational Linguistics (2014)
45. Peters, M.E., et al.: KnowBERT: knowledge-enhanced contextualized word representations. In: Proceedings of EMNLP 2019 (2019)
46. Petroni, F., et al.: Language models as knowledge bases? In: Inui, K., Jiang, J., Ng, V., Wan, X. (eds.) Proceedings of the 2019 Conference on Empirical Methods in Natural Language Processing and the 9th International Joint Conference on Natural Language Processing (EMNLP-IJCNLP), Hong Kong, China, pp. 2463–2473. Association for Computational Linguistics (2019)
47. Pound, J., Mika, P., Zaragoza, H.: Ad-hoc object retrieval in the web of data. In: Rappa, M., Jones, P., Freire, J., Chakrabarti, S. (eds.) Proceedings of the 19th International Conference on World Wide Web, WWW 2010, Raleigh, North Carolina, USA, 26–30 April 2010, pp. 771–780. ACM (2010)
48. Radford, A., Jeffrey, W., Child, R., Luan, D., Amodei, D., Sutskever, I.: Language models are unsupervised multitask learners. Technical report, OpenAI Technical Report (2019)
49. Raffel, C., et al.: Exploring the limits of transfer learning with a unified text-to-text transformer. J. Mach. Learn. Res. **21**, 140:1–140:67 (2020)
50. Ruffinelli, D., Broscheit, S., Gemulla, R.: You CAN teach an old dog new tricks! On training knowledge graph embeddings. In: 8th International Conference on Learning Representations, ICLR 2020, Addis Ababa, Ethiopia, 26–30 April 2020. OpenReview.net (2020)
51. Sequeda, J., Lassila, O.: Designing and Building Enterprise Knowledge Graphs. Synthesis Lectures on Data, Semantics, and Knowledge. Morgan & Claypool Publishers (2021)

52. Sevgili, Ö., Shelmanov, A., Arkhipov, M., Panchenko, A., Biemann, C.: Neural entity linking: a survey of models based on deep learning. Semant. Web **13**(3), 527–570 (2022)

53. Speer, R., Chin, J., Havasi, C.: Conceptnet 5.5: an open multilingual graph of general knowledge. In: Singh, S.P., Markovitch, S. (eds.) Proceedings of the Thirty-First AAAI Conference on Artificial Intelligence, 4–9 February 2017, San Francisco, California, USA, pp. 4444–4451. AAAI Press (2017)

54. Sun, Y., Wang, S., Li, Y., et al.: Ernie: enhanced representation through knowledge integration. In: AAAI (2019)

55. Thoppilan, R., et al. LaMDA: language models for dialog applications. ArXiv preprint, abs/2201.08239 (2022)

56. Tsaneva, S., Dessì, D., Osborne, F., Sabou, M.: Knowledge graph validation by integrating LLMs and human-in-the-loop. Inf. Process. Manage. **62**(5), 104145 (2025)

57. Vaswani, A., et al.: Attention is all you need. In: Guyon, I., et al. (eds.) Advances in Neural Information Processing Systems 30: Annual Conference on Neural Information Processing Systems 2017, 4–9 December 2017, Long Beach, CA, USA, pp. 5998–6008 (2017)

58. Wang, J., et al.: Learning to plan for retrieval-augmented large language models from knowledge graphs. arXiv preprint arXiv:2406.14282 (2024)

59. Wang, M., Qiu, L., Wang, X.: A survey on knowledge graph embeddings for link prediction. Symmetry **13**(3), 485 (2021)

60. Wen, H., Chen, M., Chang, K.-W., Roth, D.: MindMap: prompting large language models with knowledge graphs for graph-of-thought reasoning. In: Proceedings of the 62nd Annual Meeting of the Association for Computational Linguistics (ACL). Association for Computational Linguistics (2024)

61. [First Name] Yang and Others. Fact-aware generation in large language models (2024). https://example.org/yang2024factaware. Extended version available online

62. Zeng, X., Xinqi, T., Liu, Y., Xiangzheng, F., Yansen, S.: Toward better drug discovery with knowledge graph. Curr. Opin. Struct. Biol. **72**, 114–126 (2022)

63. Zhang, H., Guo, Z., et al.: Kg-enhanced reasoning in large language models. In: Proceedings of ACL 2024 (2024)

64. Zhang, Y., Chen, Z., Guo, L., Xu, Y., Zhang, W., Chen, H.: Making large language models perform better in knowledge graph completion. arXiv preprint arXiv:2310.06671 (2023)

65. Zhang, Y., Wang, X., Liang, J., Xia, S., Chen, L., Xiao, Y.: Chain-of-knowledge: integrating knowledge reasoning into large language models by learning from knowledge graphs. arXiv preprint arXiv:2407.00653 (2024)

66. Zhao, W.X., et al.: A survey of large language models. ArXiv preprint, abs/2303.18223 (2023)

67. Zhao, Z., Luo, X., Chen, M., Ma, L.: A survey of knowledge graph construction using machine learning. Comput. Model. Eng. Sci. **139**(1), 225–257 (2024)

68. Zhu, Y., et al.: LLMs for knowledge graph construction and reasoning: recent capabilities and future opportunities. arXiv preprint arXiv:2305.13168 (2023)

Recent Advances and Future Directions in Literature-Based Discovery

Andrej Kastrin[1]([✉])(iD), Bojan Cestnik[2,3](iD), and Nada Lavrač[2](iD)

[1] University of Ljubljana, Ljubljana, Slovenia
andrej.kastrin@mf.uni-lj.si
[2] Jožef Stefan Institute, Ljubljana, Slovenia
bojan.cestnik@temida.si, nada.lavrac@ijs.si
[3] Temida d.o.o, Ljubljana, Slovenia

Abstract. The explosive growth of scientific publications has created an urgent need for automated methods that facilitate knowledge synthesis and hypothesis generation. Literature-based discovery (LBD) addresses this challenge by uncovering previously unknown associations between disparate domains. This article surveys recent methodological advances in LBD, focusing on developments from 2000 to the present. We review progress in three key areas: knowledge graph construction, deep learning approaches, and the integration of pre-trained and large language models (LLMs). While LBD has made notable progress, several fundamental challenges remain unresolved, particularly concerning scalability, reliance on structured data, and the need for extensive manual curation. By examining ongoing advances and outlining promising future directions, this survey underscores the transformative role of LLMs in enhancing LBD and aims to support researchers and practitioners in harnessing these technologies to accelerate scientific innovation.

Keywords: Artificial intelligence · Natural language processing · Computational scientific discovery · Literature-based discovery

1 Introduction

The explosive growth of research publications across various scientific disciplines has led to an overwhelming volume of knowledge, ranging from research articles and monographs to preprints and conference proceedings [4]. This proliferation has made it increasingly difficult for researchers to effectively locate, interpret, and synthesize relevant knowledge. As a result, staying current within one's field becomes more challenging, and the risk of missing important findings or inadvertently duplicating existing work rises significantly. Furthermore, the increasing complexity and interdisciplinarity of research further complicate the task of integrating knowledge from multiple sources, and much of the information remains siloed, underutilized, or disconnected. These challenges have led to a rising interest in developing automated methods, particularly those based on

© The Author(s), under exclusive license to Springer Nature Switzerland AG 2026
M. van Leeuwen and J. Vreeken (Eds.): Arno Siebes Festschrift, LNCS 16067, pp. 198–210, 2026.
https://doi.org/10.1007/978-3-032-03028-3_12

natural language processing (NLP), to support hypothesis generation and the discovery of novel scientific insights.

A promising approach to address this problem is literature-based discovery (LBD). LBD, originally introduced by Swanson [29] in the mid-1980s, is an approach designed to generate novel research hypotheses by revealing previously overlooked associations between two complementary and non-interactive sets of scientific literatures. It emerged as a response to the growing difficulty of staying abreast of developments across disparate fields and remains a valuable methodology in the face of ever-expanding scholarly output.

The primary motivation of this article is to provide an overview of current methodological challenges in LBD, survey recent scientific advances, and identify future research directions that align LBD with emerging trends in AI and more broadly computational scientific discovery. We limit the scope to the period between 2000 and early 2025, focusing exclusively on state-of-the-art approaches, as earlier methods have already been comprehensively covered in previous surveys [12, 28, 32].

The article is organized as follows. Section 2 presents the necessary preliminaries and a concise overview of LBD research. Recent advances in LBD methodologies are examined in Sect. 3, followed by a discussion of future research directions in Sect. 4. The article concludes with a synthesis of key findings in Sect. 5.

2   Preliminaries and Background

LBD is a subfield of artificial intelligence (AI) at the intersection of information retrieval, NLP and computational scientific discovery, which is dedicated to automating the scientific discovery process. The early Swanson's approach to LBD can be formalized using a generic ABC model (Fig. 1) that considers two independent literature sets, A and C [30]. In this model, a represents a source concept, c is a target concept, and b serves as a bridge or intermediate concept that connects the two. The key idea is that if a is associated with b in one body of literature and b is associated with c in another—yet a and c have not been directly linked in any publication—there may be a novel, undiscovered relationship between a and c worth exploring.

A seminal example of this model is Swanson's [29] groundbreaking discovery linking dietary fish oil (a) to Raynaud's disease (c). He found that Raynaud's disease was associated with reduced blood viscosity in one set of articles, while another set linked high blood viscosity to fish oil. Although no studies at the time had made a direct connection between Raynaud's disease and fish oil, Swanson's hypothesis suggested a new therapeutic use for fish oil, which was later confirmed by clinical research [8].

In general, LBD encompasses two tasks: hypothesis validation and hypothesis generation, which correspond to closed and open discovery modes, respectively. In closed discovery, the process starts with two known elements—a starting concept (a) and a target concept (c)—and seeks to validate or elaborate the relationship between them by identifying intermediate mechanisms (b) that connect a and c. Conversely, in open discovery, the process begins with a starting

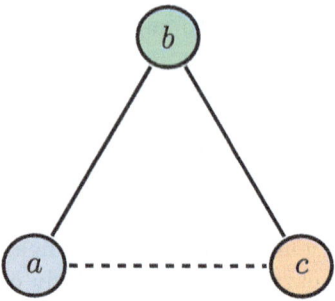

Fig. 1. Swanson's ABC model of discovery. When a is related to b, and b is related to c, it suggests the possibility of an undiscovered indirect relationship between a and c.

concept (a) and aims to uncover previously unrecognized intermediate concepts (b) and target concepts (c) that could suggest novel research hypotheses.

Despite its pivotal importance, the ABC model exhibits critical limitations that severely constrain the broader applicability of LBD. First, scalability remains a pressing challenge. Traditional LBD systems were developed for relatively small, curated datasets and are poorly suited to handle the exponential growth of biomedical publications [32]. Effectively managing large-scale, heterogeneous corpora demands advanced computational capabilities and methodological innovations that classical LBD frameworks were not originally designed to support.

Second, the heavy reliance on structured data sources represents a major constraint. LBD approaches have historically depended on controlled vocabularies and ontologies, such as Unified Medical Language System (UMLS) [3], which facilitate computational access but simultaneously narrow the scope of discovery to well-represented areas [12]. Consequently, current LBD systems often exhibit limited flexibility when extracting knowledge directly from unstructured or semi-structured texts, which account for a substantial portion of the scientific literature.

Third, the reliance on extensive manual curation and expert intervention remains a substantial barrier to progress in LBD. Traditional LBD workflows necessitate expert involvement at multiple stages, including hypothesis validation, result refinement, and relevance assessment [28]. This dependence not only slows the overall discovery process but also poses significant challenges to achieving the scalability and reproducibility required for the broader application of LBD tools.

The landscape of LBD is evolving rapidly, but a comprehensive approach to tackling these challenges is essential for realizing its full potential in biomedical research and beyond.

3   Recent Advances

The field of LBD has seen notable progress in recent years, driven largely by advancements in machine learning, text mining, and statistical analysis. Research efforts have increasingly harnessed these technologies to develop more effective and sophisticated LBD systems. This section reviews three major directions contributing to the recent evolution of LBD: knowledge graphs (KGs), deep learning (DL), and language models (LMs).

3.1   Knowledge Graphs

KGs have emerged as a pivotal technology in NLP, offering a structured and scalable approach to organizing scientific knowledge. By representing information as networks of entities and their relationships, KGs enable graph-based reasoning and facilitate the identification of implicit associations across disparate literature sources.

Formally, KGs are defined as $G = (V, E)$, where V represents the set of vertices (nodes) and E represents edges (links). Relationships in the graph are often modeled as triples (h, r, t), where h (head) and t (tail) are nodes and r is the relation connecting them.

Their construction typically follows two principal methodologies: (i) co-occurrence modeling, where links between entities are established if they co-appear within the same article; and (ii) explicit relation extraction, where semantic relationships are directly identified using specialized NLP tools such as SemRep [26]. Co-occurrence models are widely adopted due to their simplicity and scalability, while relation extraction methods provide greater precision and richer semantic information.[1] In particular, co-occurrence-based approaches have gained popularity in LBD systems owing to their ease of implementation [12, 28]. Recent approaches also enable the direct construction of KGs based on predications (subject-relation-object triples) extracted from sources such as PubMed abstracts. Resources such as the UMLS and Open Biomedical Ontologies (OBO) offer rich terminological frameworks that enhance the integration and cross-referencing of knowledge.

We approach LBD by formulating it as a knowledge graph completion (KGC) task. KGC techniques aim to predict missing information in graphs, either by discovering new edges (link prediction) or by identifying missing nodes (node prediction). Depending on the method used to construct the KG, the elements h, r, and t (as previously defined) may differ: in co-occurrence-based graphs, all three components often represent concepts or terms, whereas in relational databases, h and t denote entities and r represents a predicate, such as TREATS.

---

[1] Co-occurrence refers to the statistical tendency of two terms or concepts to appear together in text (e.g., *fever* and *infection*), without implying a specific semantic or causal relationship. In contrast, a semantic relation denotes a defined and meaningful connection between terms, such as a taxonomic link (e.g., *influenza* as a type of *viral infection*), regardless of how frequently they co-occur.

Examples include structures like "Fish oil" → "Blood viscosity" → "Raynaud's disease" for co-occurrence graphs, or "Fish oil" → TREATS → "Raynaud's disease" for relational graphs.

Two main approaches to KGC are usually employed: (i) evaluating the plausibility of candidate triples (h, r, t) by assigning a predictive score, and (ii) inferring missing elements by submitting incomplete triples, such as (h, r), (h, t), or (r, t), and predicting the missing component (i.e., predicting t given (h, r), r given (h, t), or h given (r, t)).

3.2 Deep Learning

In contrast to traditional machine learning methods that rely on features explicitly constructed using domain knowledge, DL uses specialized and deep architectures to extract meaningful features from unstructured input, can automatically learn from simple inputs, and extracts task-specific representations of structures.

Crichton et al. [5] provided compelling evidence that neural network models are highly effective for advancing LBD. The authors built upon a multilayer perceptron (MLP) framework designed for both closed and open discovery tasks, achieving state-of-the-art performance on the PubTator and BioGRID datasets. Their approach begins by generating input representations through node embeddings using the Large-scale Information Network Embedding (LINE) algorithm [31], followed by various strategies for combining the embeddings of nodes along a discovery path—structured according to Swanson's ABC model—to construct the input for the neural model.

In closed discovery, the first approach uses a neural model to assign scores to individual $A-B$ and $B-C$ links, which are then aggregated to evaluate the full $A-B-C$ path. The second approach combines the embeddings of A, B, and C into a single input vector, allowing the model to predict a score for the entire path directly, thus removing the need for an explicit aggregation step. In open discovery, the first method similarly scores $A-B$ and $B-C$ links and aggregates them, but additionally uses an accumulator function to integrate multiple paths leading to the same C. The second method employs a convolutional neural network (CNN) that processes stacked embeddings of multiple $A-B-C$ paths, producing a unified score for each $A-C$ pair without relying on separate aggregation or accumulation functions. (Unlike conventional convolutional neural network (CNN) applications where images are used, the input here is a pseudo-image created by stacking vectorized $A-B-C$ paths.)

While Crichton et al. [5] relied on embedding representations for all concepts as model inputs, their method required users to manually construct all possible hypothesis triples prior to evaluation, a process that is both time-consuming and reliant on substantial domain expertise. Addressing this limitation, Cuffy et al. [6] introduced a closed discovery framework that automates the ranking of potential linking B terms for a given A and C pair using a single forward propagation step through the DL model. This approach eliminates the need to generate all $A-B-C$ triples *a priori*, thereby reducing the dependency on domain-specific knowledge and significantly streamlining the LBD workflow.

Cuffy et al. [7] introduced a further advancement by reformulating the LBD task as the prediction of implicit concept embeddings rather than direct relationship scoring. Instead of classifying triples, their model predicts the embedding of the linking concept (B) given the starting (A) and target (C) concepts. By comparing predicted embeddings against all candidate concepts, the MLP model identifies plausible intermediates, demonstrating its effectiveness in systematic knowledge discovery replication.

Beyond general-purpose LBD tasks, DL has been effectively applied in domain-specific applications, such as drug repurposing. Zhu et al. [39] introduced a BioBERT-based model enhanced with entity-aware attention mechanisms for drug-drug interaction extraction, while Gupta et al. [10] utilized an NSGA-III-based CNN architecture to optimize biomedical search engines. Rather et al. [25] further showcased DL's capacity to uncover latent biomedical relationships through word2vec-based embeddings. Taken together, these studies demonstrate the transformative potential of DL for LBD, facilitating more nuanced knowledge representation and discovery.

3.3   Language Models

LMs are nowadays regarded as fundamental components of NLP, tasked with estimating the probability distributions of linguistic units—such as words, phrases, or sentences—based on their contextual surroundings. The evolution of LMs can be delineated into several distinct stages: beginning with statistical language models (SLMs), progressing through neural language models (NLMs), advancing to pre-trained language models (PLMs), and culminating in the emergence of large language models (LLMs). SLMs utilize basic probabilistic frameworks to model word sequences (e.g., n-grams), whereas NLMs employ neural networks to capture complex syntactic and semantic patterns (e.g., RNNs, LSTMs, transformers). PLMs leverage large-scale textual corpora and self-supervised learning to encode general linguistic structures and knowledge (e.g., Bidirectional Encoder Representations from Transformers (BERT), Generative Pretrained Transformer (GPT)).

Here, we focus specifically on how recent developments in PLMs and LLMs have been integrated into LBD pipelines. While both PLMs and LLMs are trained on large corpora using self-supervised methods, LLMs represent a significant advancement in terms of model size, training data scale, and architectural complexity. Building upon the foundation established by PLMs, LLMs offer improved generalization, greater expressivity, enhanced contextual understanding, adaptability, and zero-shot reasoning capabilities—making them particularly well-suited for advanced LBD tasks. A comprehensive overview of the historical development of LMs is provided in recent surveys by Wang et al. [34] and Annepaka et al. [1].

PLMs have elevated the quality and scope of LBD by integrating deep contextual understanding into NLP pipelines. Used both as powerful preprocessing tools and as core components in downstream tasks such as named entity recognition and relation extraction, PLMs have enabled more accurate and scalable discovery workflows. For example, in our LBD approach to drug repurposing for Covid-19 [38], we employed BERT[2] as a preprocessing tool to generate an accurate subset of semantic triples, which were then used to construct a KG. KGC algorithms were subsequently applied to this graph to predict potential drug repurposing candidates.

Compared to PLMs, LLMs exhibit remarkable adaptability, with recent empirical results indicating strong potential for their use as a general-purpose tool to support scientific reasoning [13]. A growing body of evidence reveals a broad range of promising capabilities of LLM relevant to the scientific process, including the coherent integration of diverse knowledge concepts, the critical evaluation of existing studies, the generation of scientific hypotheses, and the identification of research gaps within scientific literature [21]. State-of-the-art LLMs, such as BioGPT [20] and SciGLM [37] are trained on domain-specific corpora like PubMed and arXiv and are particularly effective at literature retrieval, document summarization, and question answering. They facilitate more efficient access to scientific information by identifying relevant publications, extracting key insights, and synthesizing knowledge across documents.

Specifically, building on the improved reasoning capabilities of LMs, the LBD community has begun developing methods that incorporate richer contextual information to address the limitations of traditional approaches, which are primarily based on the ABC model. Classical LBD techniques often fail to capture the nuanced contextual cues considered by human scientists during the ideation process and are largely restricted to predicting pairwise relationships between isolated concepts [13]. To overcome these constraints, Wang et al. [33] introduced a novel framework, SciMON, which grounds LBD in natural language contexts, thereby narrowing the generative space in a more controlled and meaningful way. SciMON optimizes novel research hypotheses by iteratively refining idea suggestions derived from published literature until sufficient novelty is achieved. Unlike traditional models that merely predict conceptual links, SciMON generates complete sentences as outputs, offering a more nuanced and contextually rich representation of scientific knowledge. The authors report that it produces ideas that are both more original and exhibit greater conceptual depth than those generated by GPT-4.

A long-standing limitation of LBD has been its restriction to the biomedical domain, primarily due to the widespread availability of the PubMed database and auxiliary knowledge resources (e.g., UMLS vocabularies [3], SemMedDB repository [16], PubTator annotations [35]), which are freely available and optimized for computational access and analysis. However, LLM-powered LBD may have a much broader scope of applicability. In particular, Yang et al. [36] showed that the majority of published hypotheses in the social sciences can be structured in a manner compatible with the LBD framework.

---

[2] Due to its success in general NLP tasks, BERT has been adapted to various specialized domains, including biomedicine, resulting in models such as BioBERT [18], ClinicalBERT [14], PubMedBERT [9], and COVID-Twitter-BERT [23].

In summary, recent advances in KGs, DL techniques, and LM development have significantly expanded the capabilities of LBD. Table 1 highlights the principal characteristics of the described approaches. Nevertheless, several key challenges remain, which are discussed in the following section.

Table 1. Summary of key strengths and limitations of recent approaches in LBD.

Approach	Strengths	Limitations
KGs	Captures complex, heterogeneous associations; enables context-driven subgraph creation.	Requires high-quality semantic annotation; extensive filtering often needed.
LMs	Offers potential for explainable AI; capable of processing heterogeneous, cross-domain data.	Still emerging; scalability and validation challenges.
DL	Outperforms traditional baselines, especially when input representations are well-optimized.	Interpretability remains a challenge; generalizability may be constrained by data representation.

4 Future Directions

Although LBD has made significant advances over the past five years, numerous open challenges remain to be addressed. The following discussion outlines several key areas of ongoing work, reflecting the current focus of our research efforts; however, the list is not intended to be exhaustive.

4.1 Advancing Interpretability

Interpretability remains one of the principal challenges associated with the application of DL techniques in science [24]. Ensuring interpretability in LBD is not simply an auxiliary feature; it is foundational. While DL approaches offer considerable potential for enhancing hypothesis generation from large corpora, their inherent "black-box" nature continues to present significant obstacles for scientific domains where transparent reasoning processes are essential. In particular, many LBD methods, especially those rooted in Swanson's ABC model, focus primarily on hypothesis generation but often lack mechanisms for explaining the reasoning behind the generated hypotheses. While DL systems excel in extracting patterns from literature, they frequently fall short of providing understandable explanations, which symbolic systems have historically offered [2].

Traditional strategies, such as employing attention mechanisms or inspecting model coefficients, offer partial solutions by highlighting feature importance or visualizing internal representations [22]. However, these approaches often lack a structured reasoning component and thus fall short of delivering full scientific explanatory power. A promising direction is the integration of neuro-symbolic AI into LBD methodologies. The neuro-symbolic approach aims to combine the pattern recognition capabilities of neural networks with the explicit reasoning structures of symbolic AI [2,27]. This integration enables models not only to learn from data but also to reason in ways that are inherently interpretable and grounded in logical principles. Neuro-symbolic approaches have already been successfully applied to various NLP tasks [11].

4.2   Augmenting Data Resources

One of the principal limitations of current LBD applications lies in their restricted use of data resources. Most existing approaches rely primarily on Pub-Med, often limiting their textual input to article abstracts rather than utilizing the full texts. While abstracts offer a concise summary of findings, they frequently omit critical contextual relationships that could be valuable for complex hypothesis generation, particularly for LMs. Expanding beyond the biomedical domain to include full-text articles and additional knowledge bases presents a significant opportunity for advancing LBD. In particular, new bibliographic databases such as Semantic Scholar [19] have emerged as valuable resources. These platforms aggregate extensive metadata, citation networks, and, in some cases, full-text content across a wide range of scientific disciplines, offering richer semantic contexts for discovery processes.

In addition, auxiliary knowledge resources, such as the UMLS for the biomedical domain, are of significant importance, particularly during the preprocessing stages of LBD (e.g., guiding the extraction of knowledge concepts and the computation of predicates). Although widely integrated into LBD applications for its standardized vocabularies and extensive concept mappings, the use of UMLS is not without limitations. Issues such as term ambiguity and incomplete concept coverage can substantially impact the performance of downstream tasks. For instance, a significant portion of errors in tools like SemRep stem from difficulties in correctly identifying and normalizing biomedical entities using UMLS, accounting for up to 27% of errors in some evaluations [15]. FurthermoreIn addition, auxiliary knowledge resources, such as the UMLS for the biomedical domain, are of significant importance, particularly during the preprocessing stages of LBD (e.g., guiding the extraction of knowledge concepts and the computation of predicates). Although widely integrated into LBD applications for its standardized vocabularies and extensive concept mappings, the use of UMLS is not without limitations. Issues such as term ambiguity and incomplete concept coverage can substantially impact the performance of downstream tasks.

For instance, a significant portion of errors in tools like SemRep stem from difficulties in correctly identifying and normalizing biomedical entities using UMLS, accounting for up to 27% of errors in some evaluations [15]. Finally, to the best of our knowledge at the time of writing, no comparably well-developed knowledge resource exists outside the life sciences domain. In our experience, the limited adoption of LBD beyond biomedicine is largely due to the greater terminological diversity and, in particular, the absence of standardized ontologies in the humanities and social sciences., to the best of our knowledge at the time of writing, no comparably well-developed knowledge resource exists outside the life sciences domain. In our experience, the limited adoption of LBD beyond biomedicine is largely due to the greater terminological diversity and, in particular, the absence of standardized ontologies in the humanities and social sciences.

4.3   Refining Benchmarking Practices

Knuth's [17] concept of literate programming, which emphasizes that computer programs should be readable and understandable by humans, closely aligns with open science initiatives that stress the importance of standardized practices and tools to ensure research outputs are independently verifiable and can support further scientific progress.

Following the principles of open science, we initiated a project aimed at promoting reproducibility within the field of LBD. Existing LBD approaches and results often remain difficult to replicate due to the lack of access to original datasets and unresolved programming dependencies. These limitations pose significant barriers to both the theoretical understanding and practical reuse of previously published methods. To address this gap, we have made publicly available benchmark datasets, replicable LBD case studies, and a collection of interactive Jupyter Notebooks that transparently document each step of the LBD pipeline, including data acquisition, text preprocessing, hypothesis discovery, and evaluation. Furthermore, we provide the LBD community with access to standardized benchmark datasets and prototypical LBD techniques presented through dockerized Jupyter environments, thereby greatly simplifying replication and extension. All associated materials are openly accessible at https://github.com/akastrin/ida2025lbd.

5   Conclusion

This survey has reviewed the evolution of LBD over the past five years. We discussed the growing role of KGs, advances in DL methodologies, and the transformative impact of PLMs and LLMs on hypothesis generation and scientific reasoning.

Rapid advances in AI, particularly in the development of LLMs, are reshaping the scientific landscape at an unprecedented pace. These developments open up significant opportunities for treating scientific corpora as dynamic knowledge bases from which novel insights, hypotheses, and ideas can be systematically uncovered. Despite this progress, several fundamental challenges remain unresolved in LBD, notably issues related to scalability, dependence on structured data, the need for extensive manual curation, and the limited interpretability of current DL approaches.

Recent trends in neuro-symbolic AI suggest promising avenues for enhancing both the accuracy and explainability of LBD systems. By combining the strengths of DL with the reasoning capabilities of symbolic methods, these hybrid approaches aim to deliver more transparent and trustworthy discoveries, thereby enabling broader domain applicability of LBD beyond the biomedical sciences.

Acknowledgments. The authors acknowledge the financial support from the Slovenian Research and Innovation Agency through the Knowledge Technologies (Grant No. P2-0103), and Methodology for Data Analysis in Medical Sciences (Grant No. P3-0154) core research projects, as well as Embeddings-Based Techniques for Media Monitoring Applications (Grant No. L2-50070) research project.

Disclosure of Interests. The authors have no competing interests to declare that are relevant to the content of this article.

References

1. Annepaka, Y., Pakray, P.: Large language models: a survey of their development, capabilities, and applications. Knowl. Inf. Syst. **67**(3), 2967–3022 (2025). https://doi.org/10.1007/s10115-024-02310-4
2. Bhuyan, B.P., Ramdane-Cherif, A., Tomar, R., Singh, T.P.: Neuro-symbolic artificial intelligence: a survey. Neural Comput. Appl. **36**(21), 12809–12844 (2024). https://doi.org/10.1007/s00521-024-09960-z
3. Bodenreider, O.: The unified medical language system (UMLS): integrating biomedical terminology. Nucleic Acids Res. **32**(Database issue), D267–D270 (2004). https://doi.org/10.1093/nar/gkh061
4. Bornmann, L., Haunschild, R., Mutz, R.: Growth rates of modern science: a latent piecewise growth curve approach to model publication numbers from established and new literature databases. Humanit. Soc. Sci. Commun. **8**(1), 224 (2021). https://doi.org/10.1057/s41599-021-00903-w
5. Crichton, G., Baker, S., Guo, Y., Korhonen, A.: Neural networks for open and closed literature-based discovery. PLoS ONE **15**(5), e0232891 (2020). https://doi.org/10.1371/journal.pone.0232891
6. Cuffy, C., McInnes, B.T.: Exploring a deep learning neural architecture for closed literature-based discovery. J. Biomed. Inf. **143**, 104362 (2023). https://doi.org/10.1016/j.jbi.2023.104362
7. Cuffy, C., McInnes, B.T.: Predicting implicit concept embeddings for singular relationship discovery replication of closed literature-based discovery. Front. Res. Metrics Anal. **10**, 1509502 (2025). https://doi.org/10.3389/frma.2025.1509502
8. DiGiacomo, R.A., Kremer, J.M., Shah, D.M.: Fish-oil dietary supplementation in patients with Raynaud's phenomenon: a double-blind, controlled, prospective study. Am. J. Med. **86**(2), 158–164 (1989). https://doi.org/10.1016/0002-9343(89)90261-1

9. Gu, Y., et al.: Domain-specific language model pretraining for biomedical natural language processing. ACM Trans. Comput. Healthc. **3**(1), 2 (2021). https://doi.org/10.1145/3458754
10. Gupta, M., Kumar, N., Singh, B.K., Gupta, N.: NSGA-III-based deep-learning model for biomedical search engines. Math. Probl. Eng. **2021**(1), 9935862 (2021). https://doi.org/10.1155/2021/9935862
11. Hamilton, K., Nayak, A., Božić, B., Longo, L.: Is neuro-symbolic AI meeting its promises in natural language processing? A structured review. Seman. Web **15**(4), 1265–1306 (2024). https://doi.org/10.3233/SW-223228
12. Henry, S., McInnes, B.T.: Literature based discovery: models, methods, and trends. J. Biomed. Inf. **74**, 20–32 (2017). https://doi.org/10.1016/j.jbi.2017.08.011
13. Hope, T., Downey, D., Weld, D.S., Etzioni, O., Horvitz, E.: A computational inflection for scientific discovery. Commun. ACM **66**(8), 62–73 (2023). https://doi.org/10.1145/3576896
14. Huang, K., Altosaar, J., Ranganath, R.: ClinicalBERT: modeling clinical notes and predicting hospital readmission. arXiv (2020). https://doi.org/10.48550/arXiv.1904.05342
15. Kilicoglu, H., Rosemblat, G., Fiszman, M., Shin, D.: Broad-coverage biomedical relation extraction with SemRep. BMC Bioinformatics **21**(1), 188 (2020). https://doi.org/10.1186/s12859-020-3517-7
16. Kilicoglu, H., Shin, D., Fiszman, M., Rosemblat, G., Rindflesch, T.C.: SemMedDB: a PubMed-scale repository of biomedical semantic predications. Bioinformatics **28**(23), 3158–3160 (2012). https://doi.org/10.1093/bioinformatics/bts591
17. Knuth, D.E.: Literate programming. Comput. J. **27**(2), 97–111 (1984)
18. Lee, J., et al.: BioBERT: a pre-trained biomedical language representation model for biomedical text mining. Bioinformatics **36**(4), 1234–1240 (2020). https://doi.org/10.1093/bioinformatics/btz682
19. Lo, K., Wang, L.L., Neumann, M., Kinney, R., Weld, D.: S2ORC: the semantic scholar open research corpus. In: Jurafsky, D., Chai, J., Schluter, N., Tetreault, J. (eds.) Proceedings of the 58th Annual Meeting of the Association for Computational Linguistics, pp. 4969–4983. Association for Computational Linguistics (2020). https://doi.org/10.18653/v1/2020.acl-main.447
20. Luo, R., et al.: BioGPT: generative pre-trained transformer for biomedical text generation and mining. Briefings Bioinform. **23**(6), bbac409 (2022). https://doi.org/10.1093/bib/bbac409
21. Luo, Z., Yang, Z., Xu, Z., Yang, W., Du, X.: LLM4SR: a survey on large language models for scientific research. arXiv (2025). https://doi.org/10.48550/arXiv.2501.04306
22. Mersha, M., Lam, K., Wood, J., AlShami, A.K., Kalita, J.: Explainable artificial intelligence: a survey of needs, techniques, applications, and future direction. Neurocomputing **599**, 128111 (2024). https://doi.org/10.1016/j.neucom.2024.128111
23. Müller, M., Salathé, M., Kummervold, P.E.: COVID-Twitter-BERT: a natural language processing model to analyse COVID-19 content on Twitter. Front. Artif. Intell. **6**, 1023281 (2023). https://doi.org/10.3389/frai.2023.1023281
24. Murdoch, W.J., Singh, C., Kumbier, K., Abbasi-Asl, R., Yu, B.: Definitions, methods, and applications in interpretable machine learning. Proc. Natl. Acad. Sci. **116**(44), 22071–22080 (2019). https://doi.org/10.1073/pnas.1900654116
25. Rather, N.N., Patel, C.O., Khan, S.A.: Using deep learning towards biomedical knowledge discovery. Int. J. Math. Sci. Comput. **3**(2), 1–10 (2017). https://doi.org/10.5815/ijmsc.2017.02.01

26. Rindflesch, T.C., Fiszman, M.: The interaction of domain knowledge and linguistic structure in natural language processing: interpreting hypernymic propositions in biomedical text. J. Biomed. Inform. **36**(6), 462–477 (2003). https://doi.org/10.1016/j.jbi.2003.11.003

27. Sarker, M.K., Zhou, L., Eberhart, A., Hitzler, P.: Neuro-symbolic artificial intelligence: current trends. AI Commun. **34**(3), 197–209 (2021). https://doi.org/10.3233/AIC-210084

28. Sebastian, Y., Siew, E.G., Orimaye, S.O.: Learning the heterogeneous bibliographic information network for literature-based discovery. Knowl.-Based Syst. **115**, 66–79 (2017). https://doi.org/10.1016/j.knosys.2016.10.015

29. Swanson, D.R.: Fish oil, Raynaud's syndrome, and undiscovered public knowledge. Perspect. Biol. Med. **30**(1), 7–18 (1986). https://doi.org/10.1353/pbm.1986.0087

30. Swanson, D.R.: Undiscovered public knowledge. Libr. Q. **56**(2), 103–118 (1986). https://doi.org/10.1086/601720

31. Tang, J., Qu, M., Wang, M., Zhang, M., Yan, J., Mei, Q.: LINE: large-scale information network embedding. In: Gangemi, A., Leonardi, S., Panconesi, A. (eds.) Proceedings of the 24th International Conference on World Wide Web, pp. 1067–1077. International World Wide Web Conference Committee (2015). https://doi.org/10.1145/2736277.2741093

32. Thilakaratne, M., Falkner, K., Atapattu, T.: A systematic review on literature-based discovery: general overview, methodology, & statistical analysis. ACM Comput. Surv. **52**(6), 129 (2019). https://doi.org/10.1145/3365756

33. Wang, Q., Downey, D., Ji, H., Hope, T.: SciMON: scientific inspiration machines optimized for novelty. In: Ku, L.W., Martins, A., Srikumar, V. (eds.) Proceedings of the 62nd Annual Meeting of the Association for Computational Linguistics, pp. 279–299. Association for Computational Linguistics (2024). https://doi.org/10.18653/v1/2024.acl-long.18

34. Wang, Z., Chu, Z., Doan, T.V., Ni, S., Yang, M., Zhang, W.: History, development, and principles of large language models: an introductory survey. AI Ethics **5**, 1955–1971 (2025). https://doi.org/10.1007/s43681-024-00583-7

35. Wei, C.H., et al.: PubTator 3.0: An AI-powered literature resource for unlocking biomedical knowledge. Nucleic Acids Res. **52**(W1), W540–W546 (2024). https://doi.org/10.1093/nar/gkae235

36. Yang, Z., Du, X., Li, J., Zheng, J., Poria, S., Cambria, E.: Large language models for automated open-domain scientific hypotheses discovery. In: Ku, L.W., Martins, A., Srikumar, V. (eds.) Findings of the Association for Computational Linguistics: ACL 2024, pp. 13545–13565. Association for Computational Linguistics (2024). https://doi.org/10.18653/v1/2024.findings-acl.804

37. Zhang, D., et al.: SciGLM: training scientific language models with self-reflective instruction annotation and tuning. arXiv (2024). https://doi.org/10.48550/arXiv.1904.05342

38. Zhang, R., Hristovski, D., Schutte, D., Kastrin, A., Fiszman, M., Kilicoglu, H.: Drug repurposing for COVID-19 via knowledge graph completion. J. Biomed. Inform. **115**, 103696 (2021). https://doi.org/10.1016/j.jbi.2021.103696

39. Zhu, Y., Li, L., Lu, H., Zhou, A., Qin, X.: Extracting drug-drug interactions from texts with BioBERT and multiple entity-aware attentions. J. Biomed. Inform. **106**, 103451 (2020). https://doi.org/10.1016/j.jbi.2020.103451

SIEBES: A Case Study in Hallucinatory Hagiography

Johannes Fürnkranz[(✉)] [iD]

Institute for Application-Oriented Knowledge-Processing (FAW),
Johannes Kepler University, Linz, Austria
juffi@faw.jku.at

Abstract. Large Language Models (LLMs) have rapidly permeated both personal and professional domains, extending their use beyond everyday problem-solving to scientific research. In this paper, we explore a less common but culturally significant application: assistance in the writing of a contribution to a liber amicorum — an edited volume celebrating a distinguished scholar. To support this delicate task, we introduce a semi-automated System for Interpretable and Explainable Biographical Extrapolation and Synthesis (SIEBES), and report our experiences with its use in a case study. Particular attention is paid to the phenomenon of hallucinations, providing a formal definition and empirical insights. Our study highlights both the promises and perils of AI-assisted biographical writing in the context of an esteemed academic tradition. (Disclaimer: Parts of this abstract have been written by a large language model.)

Keywords: Arno Siebes · Birthday · Retirement

1 Introduction

Lately, Large Language Models (LLMs) have received a lot of attention [27], and, in fact, quickly permeated into our personal and professional lifes. They are no longer restricted to their intended use as an easily accessible pseudo-encyclopedic interactive knowlege base, typically used for solving programming challenges or homework problems, but have become an essential tool in scientific research [2,19]. They have, for example, successfully been employed for supporting and (in some cases) even automating a variety of tedious tasks, not only scholarly writing [30] but also peer reviewing [7,13], meta-reviewing [8], quality assurance [18], or searching for relevant literature [1].

In this paper, we study a rather rare but all the more important task: the writing of a contribution to a so-called *liber amicorum*, an edited volume consisting of contributions by a prominent scientist's peers, colleagues, students, or friends (these categories are not necessarily mutually exclusive), typically collected on the occasion of a landmark birthday or retirement (or both!). Prime examples of the genre include [3, 21].

© The Author(s), under exclusive license to Springer Nature Switzerland AG 2026
M. van Leeuwen and J. Vreeken (Eds.): Arno Siebes Festschrift, LNCS 16067, pp. 211–226, 2026.
https://doi.org/10.1007/978-3-032-03028-3_13

When being invited to contribute to such a volume, an author typically faces a particular challenge: on the one hand, he[1] would certainly have papers to contribute (rejects abound these days), but these will have little to do with the work of the honoree (for a notable exception we refer to [11]). While it is often very easy to see the connection of one's own work to a paper at hand in a reviewing context, it becomes much harder to actually prove this in writing. On the other hand, the prospective author may also wish to contribute something personal, maybe pepper up the contribution with an embarrassing anecdote or two, and in general demonstrate a special bond with the dedicatee. However, for achieving this goal, the poor author often lacks the necessary familiarity with his victim's general life course, as he only knows him from regular but informal meetings, in settings where one's memory is often foggy and blurred. In desperation, an author may thus be tempted to solicit help from the outside, be it co-authors who complement one's (lack of) knowledge or — more recently — large language models who may be able to fill in some gaps with well-worn platitudes.

As a first attempt for a remedy, we present in this work a *System for Interpretable and Explainable Biographical Extrapolation and Synthesis (SIEBES)*, and analyze its performance in a case study for complementing a (hypothetical) contribution to a *liber amicorum*. The main danger with such a system is that its reliance on large language models may occasionally result in a so-called *hallucination*, i.e., a plausible but factually incorrect statement. Needless to say, the inclusion of such statements into an article of praise and reverence would be devastating for the author. In our study, we will thus pay special attention to such hallucinations. In particular, we also contribute a formal definition of this elusive concept.

We start with a brief sketch of the system in Sect. 2, and then proceed to formally define hallucinations in Sect. 3. The setup for our case study will be described in Sect. 4, and the key findings can be found in Sect. 5. Finally, we discuss limitations of our approach (Sect. 6), and conclude (Sect. 7).

2   Methodology

Figure 1 shows the basic architecture of SIEBES. In the center is the human user. Obviously, she is in the loop, in fact even doubly so: one loop represents the writing of the contribution to the *liber amicorum*, where the user identifies an information gap and patches it with the responses obtained from the LLM. For obtaining this feedback, the user has to engineer a suitable prompt $\mathcal{P}$ that encourages the LLM to produce the desired response R. This is a non-trivial task, which is beyond the scope of this paper. We refer the interested reader to [20]. The second loop shows the interaction with the LLM, which, after being prompted, will return a response, which may be correct (a factual patch) or not (a hallucination).

---

[1] In the context of this work, for obvious reasons and without loss of generality, we assume that both the author and the honoree are male (unless mentioned otherwise).

The final task of the user is then to identify possible hallucinations in order to avoid providing misleading or even factually incorrect information, and to patch the obtained information into the developing article. This is, again, a difficult task, where the user has to strike a balance between the desired level of detail and the factual correctness. General statements will have a higher probability of being correct, but are naturally not well adapted to the honoree. Conversely, very concrete statements run into an increased danger of being factually incorrect. In Sect. 5, we illustrate various aspects of this danger in a case study.

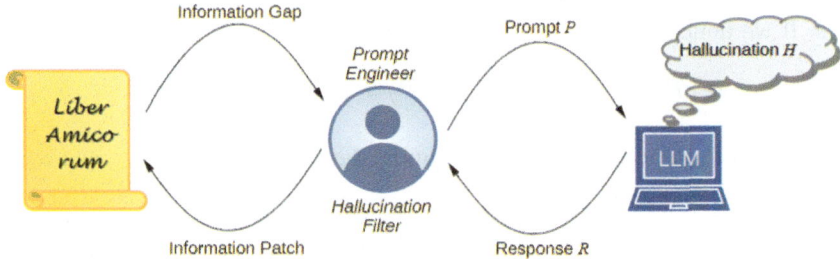

Fig. 1. The SIEBES architecture

As of now, the SIEBES system as depicted in Fig. 1 is a research prototype with a very loose integration of its integral parts. We are currently working on a full-blown implementation. The source code will be made available upon acceptance of this paper.

3   Theoretical Foundations

While the phenomenon of hallucinations in LLMs is widely known and has, naturally, received quite some attention in the recent AI literature [5,22], there are only few attempts to formally define the phenomenon. A notable exception is [14], where hallucinations are defined relative to background knowledge in the form of a knowledge graph. However, this approach makes the assumption that the knowledge graph used for checking for hallucinations is complete, which will rarely be the case in practice.

In this work, we take an alternative approach, and define hallucinations via the universal principle of compression. For a detailed treatment of this topic we refer to [24], but only mention in passing its predominant use in frequent pattern mining systems such as KRIMP [29] or DEEPKRIMP [28]. We argue that LLMs also (even if only implicitly) operate according to this principle. Obviously, the response R that an LLM should return for a given prompt $\mathcal{P}$ is the one that fits best into the context provided by $\mathcal{P}$. This informal notion of "fit" can be made precise using Kolmogorov complexity [17]. We define the *description length* of a response R to a prompt $\mathcal{P}$ as

$$DL(R, \mathcal{P}) = L(R) + L(\mathcal{P}|R) \tag{1}$$

where $L(R)$ is the length of the response in bits, and $L(\mathcal{P}|R)$ is the length of the provided prompt in the context of the given response. Note how the intuitive compression of the response in the context of the prompt is, in best Bayesian tradition, reduced to the compression of the prompt in the context of the response. Obviously, a random response R_- would not allow any reduction of the encoding length of the prompt $(L(\mathcal{P}|R_-) = L(\mathcal{P}))$, whereas an optimal response R_+ will allow to uniquely determine the prompt $(L(\mathcal{P}|R_+) = 0)$.[2] The addition of the term $L(R)$ penalizes unnecessarily verbose responses.

Our framework rests upon the assumption that an LLM returns a response that minimizes the description length (1). Note that this minimization is not unique. If we denote with

$$\mathcal{R}^*(\mathcal{P}) = \{R' \mid DL(R', \mathcal{P}) = \min_R DL(R, \mathcal{P})\} \tag{2}$$

the set of all responses with minimum description length, we therefore assume that

$$R_{LLM} \in \mathcal{R}^*(\mathcal{P}). \tag{3}$$

However, as we have argued in previous work [25], shorter explanations are not necessarily preferable over longer ones, an argument which can arguably be transferred to our current setting. We can now formally define a hallucination as follows:

Definition 1 (Hallucination). *A possible response H to a prompt $\mathcal{P}$ is called a hallucination, if it*

(i) minimizes the description length (i.e., $H \in \mathcal{R}^*(\mathcal{P})$), and
(ii) has a strictly shorter description length than any of the sensible responses $\mathcal{S}$ (i.e., $\forall S \in \mathcal{S} : S \notin \mathcal{R}^*(\mathcal{P})$).

Thus, according to this definition, hallucinations happen if and only if none of the meaningful responses have a minimum description length, or in other words, the description length of the hallucination is shorter than the description length of any of the sensible responses. Formally, this yields the following

Corollary 1. *For a hallucination H and the set of sensible responses $\mathcal{S}$ to a prompt $\mathcal{P}$, it holds that*

$$DL(H, \mathcal{P}) < \min_{S \in \mathcal{S}} DL(S, \mathcal{P}) \tag{4}$$

Informally, we may thus also say that the LLM has "shorted".

---

[2] This principle is, e.g., also at the core of the quiz show *Jeopardy!*, which has been successfully tackled by IBM's Watson system [10].

4   Experimental Setup

For conducting our case study, we need a subject who is on the one hand well respected and recognized, but about whom it is, on the other hand, hard to come across detailed biographical information from reliable sources. Moreover, the subject needs to be sufficiently old to have been part of the LLM training sets.

After some deliberation, we finally decided to select Arno Siebes. He seems ideal for our purposes because he is first of all a respected professor in computer and data science, and arguably one of the founding fathers of data mining research in Europe. In fact, in a recent survey, 67% of European data mining researchers[3] stated that they first heard about data mining in a visionary technical report, co-authored by Siebes [12].

Moreover, he is also a well-known personality, regularly appearing at key data mining events, as an organizer, program chair member, speaker, or (last but not least) treasurer, but most importantly as a central figure to vivid off-topic discussions in coffee and cigarette breaks. In analogy to well-known concepts in bibliometric network analysis such as the Morik number [4], one is tempted to define the Siebes number as the minimum path length between a person and Arno Siebes in a social network where an edge between two nodes represents the property "had a beer or a cigarette together". Empirical evidence suggests that the maximum Siebes number within the European data mining community is 2, whereas many people (including the author) believe it to be 1.

Yet it is hard to come across personal or biographical information about Siebes. Surprisingly, for a man of his reputation and stature, he does not seem to have a Wikipedia entry or maintain a private home page, and his institutional Web page[4] provides only rudimentary information. Moreover, even its international version[5] is partially composed in cryptic Dutch, which makes it hard for an author to extract key biographical facts — exactly the problem we want to study in this work.

5   Results

In this section, we report the key results of SIEBES in the task laid out in the previous section, i.e., in contributing pieces of text to an article in a *liber amicorum* for our subject, Arno Siebes. In order to explore the breadth of options, we test various input contexts and output modalities. For the underlying LLM, we primarily focused on ChatGPT-4, but ensured that our results can also hold for other architectures such as Gemini 2.5. Occasionally, we take the liberty of slightly editing, reformatting or abridging the received responses for readability, at the expense of reproducability, but without compromising their statistical significance.

---

[3] No pun intended. Because of the low sample size ($N = 1$), we used Laplace's law of succession to adjust for sampling bias [6,16].

[4] https://www.uu.nl/medewerkers/APJMSiebes, accessed on 26.4.2025.

[5] https://www.uu.nl/staff/APJMSiebes, accessed on 26.4.2025.

5.1   Vanilla

Let us start in a zero-shot setting, where nothing is known about the person to whom the article will be dedicated, beyond the immediate purpose of the article, namely "as a possible contribution to a *liber amicorum* for an eminent researcher in data mining."

A Tribute to [Researcher's Name]: A Pioneer in Data Mining and Knowledge Discovery

It is a great privilege to contribute to this liber amicorum honoring [Researcher's Name], whose pioneering work has profoundly shaped the field of data mining. Over the course of a distinguished career, [he/she/they] has illuminated the path from raw data to actionable knowledge, developing foundational algorithms and inspiring generations of researchers and practitioners alike.

Beyond [his/her/their] outstanding scientific achievements, [Researcher's Name] has exemplified the spirit of collaboration and mentorship, fostering a global community dedicated to advancing the science and ethics of data-driven discovery. [His/her/their] vision continues to resonate, not only in the academic world but across industries where data mining empowers decision-making and innovation.

It is with deep respect and admiration that I celebrate [Researcher's Name]'s enduring contributions, which will continue to influence and inspire for decades to come.

We can see that the obtained piece of text is extremely useful and broadly applicable. With minor modifications (essentially filling in the blanks with suitable names and pronouns), the text can be used as is. The author wholeheartedly and fully endorses it for our case at hand.

However, the obvious disadvantage is that the result is not tailored to a specific person, and runs the risk that a very similar text has already been in use for a different person. Plagiarism is lurking around the corner. In future work, we plan to extend SIEBES to address the problem of potential duplicates, e.g., by automatically adding watermark terms to individualize the prompts.

5.2   Factual

However, if a sufficiently prominent person is the subject of the planned article, there is a high chance that documents capturing aspects of his professional interests have already been picked up by the underlying LLM. It is then fairly trivial to modify the above task by simply specifying the name of the target person. At first, we were a bit skeptical whether SIEBES would manage to excel at this task because of the above-mentioned difficulties to access biographical details of Arno Siebes (cf. Sect. 4), but the result is, again, of quite high quality:

A Tribute to Arno Siebes: Architect of Structure and Insight in Data Mining

It is a great privilege to contribute to this liber amicorum honoring Arno Siebes, whose pioneering work has left an indelible mark on the foundations of data mining and knowledge discovery. Throughout his distinguished career, Arno has combined mathematical elegance with practical insight, advancing our understanding of probabilistic modeling, pattern mining, and the very nature of meaningful structure within complex data.

Beyond his outstanding scientific achievements, Arno has always brought a spirit of intellectual curiosity, precision, and critical thinking to every discussion — inspiring not only novel algorithms and models, but also new ways of questioning what it truly means to "discover" knowledge. His mentorship and leadership have cultivated a vibrant research community, where ideas flourish under his characteristic blend of rigor and creativity.

It is with deep respect and admiration that I celebrate Arno Siebes' enduring contributions. His work will continue to shape the landscape of data science and to inspire all who seek to find clarity and structure amid complexity.

A key advantage of this approach is that it no longer necessitates filling in blanks, so that the response can, in principle, be used as is. The author does not hesitate to admit that he could not have done a better job. On the other hand, however, the risk of hallucinations is definitely increased. For example, in the above text, we would not have put the subject's expertise in probabilistic modeling in first place, as a cursory scan of DBLP[6] has not surfaced any publications with the stem "probab" in the title or having been presented at UAI.

5.3   Personal

The factual approach delivers a useful sketch of our subject's professional achievements, which can easily be further refined by iterating over SIEBES' double feedback loop (Fig. 1). We recommend this as an exercise to the keen and interested reader. However, an article in a *liber amicorum* often has a more personal touch. SIEBES is, understandably, a bit reluctant to share personal information, be it for the necessary restraint (*"It's important to respect Professor Siebes's privacy regarding his personal life."*) or a simple lack of information in accessible sources (*"As of now, there is no publicly available information confirming whether Professor Arno Siebes is married."*). One therefore has to resort to indirect routes to work towards this goal.

One approach can be to configure SIEBES to impersonate the subject for a short interview. Conveniently, it also picks its own questions:

---

[6] https://dblp.org/pid/s/ArnoSiebes.html, accessed on 28.4.2025.

Interviewer: Professor Siebes, thank you for joining us. To begin, how did you first become interested in computer science?

"Arno Siebes": *My path into computer science began with a fascination for logic and structure. In the early days, databases were a central focus, but over time I became drawn to the patterns hidden in data — how we can uncover meaning not through hypotheses, but by letting the data speak for itself.*

Interviewer: You're known for your work with the Minimum Description Length principle. Why does that matter?

"Arno Siebes": *MDL is powerful because it formalizes the idea of simplicity in data modeling. Essentially, it tells us that the best explanation is the one that compresses the data the most. It's both elegant and practical — a kind of Occam's Razor for data.*

Interviewer: How do you feel about the explosion of data science and AI in recent years?

"Arno Siebes": *It's exciting but also challenging. We're generating more data than we can interpret meaningfully. My concern is that we sometimes chase patterns without understanding their context. Good science still requires rigor — even if hypotheses seem passè.*

Interviewer: Outside your academic work, what do you enjoy?

"Arno Siebes": *I enjoy reading broadly — history, philosophy, anything that examines how we know what we know. I've also been known to recommend books, like Seb Falk's "The Light Ages", which reminds us that the Middle Ages weren't as dark as we think.*

Interviewer: Last question: what advice would you give to students entering data science today?

"Arno Siebes": *Be curious, be skeptical, and understand the limitations of your tools. It's not just about finding patterns — it's about knowing which ones matter.*

Most of the conversation may be described as professional chit-chat, with the exception of a surprisingly concrete answer to the question on non-academic pursuits. Obviously, one must be wary and carefully scrutinize provided facts for potential hallucinations. Luckily, in this case, the concerns could be eliminated after some research.[7] However, we would certainly like to have more of this personal touch in a potential article. Asking "Arno Siebes" for his hobbies does not yield useful insights: reading (*"The Light Ages!"*), walking (*"No headphones, no screens — just thought!"*), classical music, puzzles, cooking. It becomes more interesting when SIEBES offers to provide insight into Arno's typical weekends:

---

[7] https://www.newscientist.nl/recensie/de-boekentip-van-arno-siebes/.

Saturday Morning

I like to begin slowly. No alarms. Coffee first — always black — and then a bit of reading. It might be something from The Economist, or a chapter of a book I've been working through. Lately, I've been revisiting some older texts on information theory — Shannon still has a lot to teach us.

Around mid-morning, I'll take a walk through Utrecht's older neighborhoods. There's something grounding about the architecture and the rhythm of the city on weekends. I often take a small notebook with me — just in case a thought worth capturing appears.

Saturday Afternoon

If I'm not reviewing a student's paper or preparing for a lecture, I might do a bit of exploratory programming — testing an idea for a data compression approach or an experimental visualization. Not for publication, just for curiosity.

Sometimes I'll meet a colleague or a friend for lunch — often someone who doesn't work in computer science. I think it's important to hear how non-specialists perceive data, science, and technology.

Saturday Evening

Dinner is relaxed. I may cook a simple meal — pasta with a good sauce, or perhaps something Dutch and hearty. Then I'll put on some music. If it's been a demanding week, I might reach for something ambient or meditative; if I'm in a more active mood, Bach's Art of Fugue or a Glenn Gould interpretation of the Goldberg Variations.

Sunday

Sunday mornings are for long-form thinking. Sometimes I sketch lecture notes with a pencil — yes, a real one — or write in longhand. It's not efficient, but it forces clarity.

In the afternoon, I might read a paper outside of my field — biology, linguistics, even literary theory. I find value in unfamiliar structures of thought.

Sunday evening, I try to disconnect. No email. Maybe a film. Something that lingers — not necessarily cheerful, but thoughtful. Kurosawa, perhaps, or a documentary on systems — ecological, social, or technological.

Finally, we are getting somewhere. This vivid picture of a typical "Arno Siebes" weekend is very enlightening and paints a colorful picture of his personality. Fact-checking the provided information is unfortunately hard[8] but based on our positive impression so far, we give SIEBES the benefit of the doubt here.

---

[8] In a quick Google search, its recently added "AI Overview" opined that the search term "... 'Siebes Kurosawa' likely refers to a misspelling of 'Seven Samurai' ".

5.4   Graphic

Recent LLMs do not only accept inputs provided in natural language text, but are also capable of processing inputs in different modalities. As an illustration, we used SIEBES for explaining the image shown in Fig. 2, which can be found on Arno Siebes' Web page:

This image appears to be a playful and abstract drawing. It looks like the letter "A" is used as the base shape for a character's face. Over this "A" shape, the artist added:

Wavy black lines for hair, giving it a wild, messy look.

Round glasses drawn over the face.

A simple triangular nose.

A few short lines under the nose suggesting a mustache.

Fig. 2. *A sketch of Arno (Source: https://www.uu.nl/ staff/APJMSiebes)*

The overall vibe reminds of a cartoonish scientist or professor character — maybe even a humorous take on Albert Einstein, because of the crazy hair and glasses.

This is a very useful and inspiring interpretation of the shown image. In particular the association to Albert Einstein is something that — appropriate as it is — could easily be overlooked.

5.5   Associative

As we have just seen, a popular way of expressing one's admiration of a person is to relate them to an even greater or more renowned personality. While Albert Einstein is certainly a respectable choice, Arno Siebes is quite frequently associated with a different celebrity, the well-known comic musician Alfred Matthew Yankovic, widely known under his *nom de guerre* "Weird Al" [23]. Naturally, this connection is also a topic that calls for a deeper exploration with SIEBES:

Arno Siebes and "Weird Al" Yankovic are vastly different in their careers and contributions, but if we get creative, we can find some amusing parallels:

1. ***Innovation in Their Fields:*** *Siebes is known for his innovative work in data mining and algorithmic analysis, while Weird Al revolutionized musical parody. Both take existing structures (datasets or songs) and extract new meaning from them.*
2. ***Pattern Recognition:*** *Siebes literally studies pattern mining in data, while Weird Al recognizes patterns in popular music and culture to craft his parodies. In a way, both are finding hidden insights—one in data, the other in pop songs.*

3. **Cult Followings:** *Siebes is respected in the academic world for his work on algorithms, and Weird Al has a devoted fanbase that appreciates his witty take on music. Both operate in niche spaces but have made a lasting impact.*
4. **Unconventional Creativity:** *Weird Al reinvents music with humor and wordplay, just as Siebes applies mathematical principles to discover novel insights in data. Each, in their own way, transforms the ordinary into something unexpected.*

Of course, the biggest difference is that one deals in algorithms and academia, while the other deals in accordions and comedy. But if Arno ever decided to parody data mining with a song, he might just channel his inner Weird Al!

Apparently, SIEBES has uncovered a few astonishing parallels between these two personalities that reach far beyond their undeniable physical resemblance. In particular, item 3. is corroborated by the articles in this volume. The final sentence paves the way for our next investigation.

5.6   Poetic

Personalized poetry is an always welcome contribution to a *liber amicorum*. However, alas, only few have the creativity or the sense for rhyme and rhythm that is indispensable for producing a valuable contribution of this sorts. Unfortunately, SIEBES also lacks these skills, as can be seen from this feeble attempt:

Title: "Krimp It!" (to the tune of "Beat It")

> They gave me a dataset, it was big and obscene,
> Full of useless noise, what does it mean?
> I said, "No worries, man, I got just the scheme,"
> Gonna Krimp it, gonna keep it clean!

> Just toss out the junk, find the pattern inside,
> MDL's the rule, it's my data-mining guide,
> Say goodbye to bloat, let the patterns abide,
> Just Krimp it, just Krimp it!

> (Chorus)
> Krimp it! (Krimp it!)
> Shrink that data down, yeah, trim it! (Trim it!)
> MDL's the way to win it! (Win it!)
> No redundant rows, just let 'em go,
> Gotta Krimp it!

> You got terabytes, man, it's way too much,
> You need a smaller model, gotta lose that crutch,
> Don't need noise when the patterns are clutch,
> Just Krimp it, ain't that Dutch?

5.7   Ecclesiastic

Already for the medieval clerus it was customary to write flattering and idealized biographies of their peers, yearning to preserve their good deeds for posteriority, often with the goal of promoting their canonization. *"Nihil nisi bonum"* was the principle. Such hagiographies may be regarded as an interesting special case of articles for a *liber amicorum*, and therefore deserve a separate treatment. Moreover, this will also help us to demonstrate that SIEBES is able to creatively generate a verbal and iconic image of a honoree. For the case at hand, we obtained the image in Fig. 3 that clearly captures the following idea which was also suggested by SIEBES (with some minor prompt tweaking):

Saint Arno of Compression is shown standing in a flowing robe of deep blue and gold, symbolizing knowledge and value. His long, curly hair frames a serene but intelligent face, with bright, watchful eyes behind round spectacles.

In one hand he holds a scroll tightly rolled and bound with golden cord — symbolizing compressed knowledge. In the other, he carries a small, intricate press (like a medieval book press) radiating faint light, suggesting the power of reducing vast complexity into elegant simplicity.

At his feet lies a tiny, ornate chest from which tiny scrolls spill, each bearing miniature, complex patterns — symbolizing the massive information packed into small spaces.

Fig. 3. *An iconic depiction of the subject.*

Behind him, a halo made not of simple gold, but of interwoven spirals, represents the infinite compression of ideas.

His expression is one of both mischievous cleverness and profound wisdom — the saint of those who find elegant solutions hidden within vast chaos.

Again, the result produced by SIEBES is quite satisfying, and could be added to an article without substantial modifications.

5.8   Anecdotal

Finally, we also turn to the difficult topic of humor, a well-known limitation of AIs of all sorts [9]. We test this with one of the numerous gems in our subject's treasure trove of anecdotes. When his university enforced a strict non-smoking policy in campus buildings, Arno claimed that he can still smoke in his office, as long as he closes all doors and opens the window, because he then is "topologically outside". When confronted with this conundrum, SIEBES contends:

No, you are not topologically outside the building.

In topology, spaces are considered equivalent if they can be continuously deformed into each other without tearing or gluing. The key concept here is connectedness: a space is connected if there is a continuous path between any two points within it.

Even if all the entry doors to a room are closed and all its windows are open, the room remains part of the interior of the building because it is still connected to other interior spaces through the open windows. The room and the exterior air may exchange elements (like air flow), but topologically, the room is still inside because there is no discontinuous separation from the rest of the building's structure.

If the entire building were considered as a topological space, opening windows does not redefine its boundaries — it only changes its permeability. So, from a topological perspective, you are still inside the building.

Even though SIEBES has apparently missed the humorous quality of this anecdote, the answer is nevertheless interesting, pointing out a potential loophole in Arno's argumentation. At the core of SIEBES' argument seems to be the observation that Siebes' argument only holds if all other rooms have their windows closed. Otherwise, Arno's office would be connected to the outside via its open window, and — by transitivity — to the next office with an open window. However, if the latter leaves its door open, it is clearly inside the building. Thus Arno's office is topologically connected to both the inside and the outside of the building, i.e., it effectively turns into a Klein Bottle [15]. Speculations on the ramifications of such a severe spatial deformation are, unfortunately, beyond the scope of this paper.

6   Limitations

SIEBES certainly has its limitations, of which a user should be aware. First and foremost, the natural language generation capabilities of LLMs, as impressive as they are, are still not perfect. Astute readers will be able to spot AI-generated text, if authors do not expedite great care to conceal its use. A case at hand has recently occured in the medical field, where a questionable use of LLMs was uncovered by identifying the suspicious phrase *"It is important to note that as an AI language model, I can provide a general perspective, but you should consult with medical professionals for personalized advice."* in a medical treatment of the use of nanovaccines against cancer [26, p.25].[9] Naturally, we have extra-carefully

---

[9] Cited after https://pivot-to-ai.com/2025/04/12/119-springer-cancer-treatments-book-as-an-ai-language-model/. At the time of writing this article, the book has already been retracted by the publisher.

proof-read this paper to avoid any occurrences of this phrase or similar give-aways, and we recommend potential users to do the same with their articles.

Another limitation are the computational demands of SIEBES, which may exceed the capacity of the underlying LLM technology. Depending on the intensity of her involvement, the user may have to deal with problems such as *"You've hit the Free plan limit for GPT-4o. You need GPT-4o to continue this chat because there's an attachment. Your limit resets after 1:26 AM."* These can, however, be overcome with adequate third-party funding.

7 Conclusions

In this paper, we have shown that the SIEBES framework can be used for enriching articles in a *liber amicorum* with supporting text generated by a large language model. We have investigated a variety of settings — factual and associative, prosaic and poetic, textual and graphic, sacred and profane — and found the tool to be quite useful. Naturally, the user has to be wary of potential pitfalls, most notably hallucinations of the underlying LLM. However, a careful user should be able to spot and filter the occasional problems that we have observed in our study. In future work, we plan to address the remaining limitations.

Acknowledgments. Over the years, the author has benefitted from numerous enlightening, encouraging, inspiring and always entertaining conversations with Arno Siebes. Without them, this work would not have been possible. Thank you!

References

1. Agarwal, S., et al.: LitLLMs, LLMs for literature review: are we there yet? Trans. Mach. Learn. Res. (2025). https://openreview.net/forum?id=heeJqQXKg7
2. Almarie, B., Teixeira, P.E.P., Pacheco-Barrios, K., Rossetti, C.A., Fregni, F.: Editorial – the use of large language models in science: opportunities and challenges. Principles Practice Clin. Res. **9**(1), 1–4 (2023). https://doi.org/10.21801/ppcrj.2023.91.1
3. Blockeel, H., Davis, J., Meert, W., Kimmig, A., Guns, T., Marra, G., Bekker, J., Gardik, S. (eds.): Luc 60 – Liber Amicorum presented on the occasion of Luc De Raedt's 60th birthday. DTAI Research Unit, KU Leuven (2024). https://dtai.cs.kuleuven.be/events-seminars/ldr60/liber-amicorum
4. Boulicaut, J., Plantevit, M., Robardet, C.: Local pattern detection in attributed graphs. In: Michaelis, S., Piatkowski, N., Stolpe, M. (eds.) Solving Large Scale Learning Tasks. Challenges and Algorithms, pp. 168–183. Springer (2016). https://doi.org/10.1007/978-3-319-41706-6_8
5. Bruno, A., Mazzeo, P.L., Chetouani, A., Tliba, M., Kerkouri, M.A.: Insights into classifying and mitigating LLMs' hallucinations. In: Bruno, A., Pipitone, A., Manzotti, R., Augello, A., Mazzeo, P.L., Vella, F., Chella, A. (eds.) Proceedings of the 1st Workshop on Artificial Intelligence for Perception and Artificial Consciousness (AIxPAC 2023) co-located with the 22nd International Conference of the Italian Association for Artificial Intelligence (AIxIA 2023). CEUR Workshop Proceedings, vol. 3563, pp. 50–63. CEUR-WS.org, Roma (2023). https://ceur-ws.org/Vol-3563/paper_12.pdf

6. Cestnik, B.: Estimating probabilities: a crucial task in machine learning. In: Aiello, L. (ed.) Proceedings of the 9th European Conference on Artificial Intelligence (ECAI-90), pp. 147–150. Pitman, Stockholm (1990)

7. Donker, T.: The dangers of using large language models for peer review. Lancet. Infect. Dis **23**(7), 781 (2023). https://doi.org/10.1016/S1473-3099(23)00290-6

8. Du, J., et al.: LLMs assist NLP researchers: critique paper (meta-)reviewing. In: Al-Onaizan, Y., Bansal, M., Chen, Y. (eds.) Proceedings of the 2024 Conference on Empirical Methods in Natural Language Processing (EMNLP), pp. 5081–5099. Association for Computational Linguistics, Miami (2024). https://aclanthology.org/2024.emnlp-main.292

9. Ermakova, L., Bosser, A., Miller, T., Campos, R.: CLEF 2025 JOKER lab: humour in the machine. In: Hauff, C., et al. (eds.) Proceedings of the 47th European Conference on Information Retrieval (ECIR), Part V, pp. 389–397. Springer, Lucca (2025). https://doi.org/10.1007/978-3-031-88720-8_59

10. Ferrucci, D.A. (ed.): This is Watson. IBM J. Res. Dev. **56**(3) (2012). https://doi.org/10.1147/JRD.2012.2184356. Special Issue

11. Fromont, Goethals, B.: k-Morik: mining patterns to classify cartified images of Katharina. In: Michaelis, S., Piatkowski, N., Stolpe, M. (eds.) Solving Large Scale Learning Tasks. Challenges and Algorithms – Essays Dedicated to Katharina Morik on the Occasion of Her 60th Birthday, pp. 377–385. Springer (2016). https://doi.org/10.1007/978-3-319-41706-6_21

12. Holsheimer, M., Siebes, A.: Data mining: the search for knowledge in databases. Technical Report, CS-R9406, CWI, Amsterdam, The Netherlands (1994). https://ir.cwi.nl/pub/5190/05190D.pdf

13. Hosseini, M., Horbach, S.P.J.M.: Fighting reviewer fatigue or amplifying bias? Considerations and recommendations for use of ChatGPT and other large language models in scholarly peer review. Res. Integr. Peer Rev. **8**(4) (2023). https://doi.org/10.1186/s41073-023-00133-5

14. Hron, J., et al.: Training language models on the knowledge graph: insights on hallucinations and their detectability. In: Proceedings on the 1st Conference on Language Modeling (COLM), Philadelphia, PA (2024). https://arxiv.org/pdf/2408.07852

15. Klein, F.: Ueber Riemann's Theorie der Algebraischen Functionen und ihrer Integrale - Eine Ergänzung der gewöhnlichen Darstellungen. B.G. Teubner, Leipzig (1882)

16. Laplace, P.S.: Essai philosophique sur les probabilitès. Courcier, Paris (1814)

17. Li, M., Vitányi, P.: An Introduction to Kolmogorov Complexity and Its Applications. Springer-Verlag (1993)

18. Liang, W., et al.: Monitoring AI-modified content at scale: a case study on the impact of ChatGPT on AI conference peer reviews. In: Salakhutdinov, R., Kolter, Z., Heller, K., Weller, A., Oliver, N., Scarlett, J., Berkenkamp, F. (eds.) Proceedings of the 41st International Conference on Machine Learning. Proceedings of Machine Learning Research, vol. 235, pp. 29575–29620. PMLR (21–27 Jul 2024). https://proceedings.mlr.press/v235/liang24b.html

19. Luo, Z., Yang, Z., Xu, Z., Yang, W., Du, X.: LLM4SR: a survey on large language models for scientific research (2025). https://doi.org/10.48550/ARXIV.2501.04306

20. Marvin, G., Hellen, N., Jjingo, D., Nakatumba-Nabende, J.: Prompt engineering in large language models. In: Jacob, I.J., Piramuthu, S., Falkowski-Gilski, P. (eds.) Data Intelligence and Cognitive Informatics, pp. 387–402. Springer Nature Singapore, Singapore (2024). https://doi.org/10.1007/978-981-99-7962-2_30

21. Michaelis, S., Piatkowski, N., Stolpe, M. (eds.): Solving Large Scale Learning Tasks. Challenges and Algorithms - Essays Dedicated to Katharina Morik on the Occasion of Her 60th Birthday, Lecture Notes in Computer Science, vol. 9580. Springer (2016). https://doi.org/10.1007/978-3-319-41706-6

22. Perkovic, G., Drobnjak, A., Boticki, I.: Hallucinations in LLMs: understanding and addressing challenges. In: Babic, S., et al. (eds.) Proceedings of the 47th MIPRO ICT and Electronics Convention, pp. 2084–2088. IEEE, Opatija (2024). https://doi.org/10.1109/MIPRO60963.2024.10569238

23. Rabin, N., Yankovic, A.: Weird Al: The Book. Abrams Books (2012)

24. Siebes, A.: Compression is all you need (2026). In preparation

25. Stecher, J., Janssen, F., Fürnkranz, J.: Shorter rules are better, aren't they? In: Calders, T., Ceci, M., Malerba, D. (eds.) Proceedings of the 19th International Conference on Discovery Science (DS-16), pp. 279–294. Springer-Verlag (2016). https://doi.org/10.1007/978-3-319-46307-0_18

26. Thorab, N.: Advanced Nanovaccines for Cancer Immunotherapy - Harnessing Nanotechnology for Anti-Cancer Immunity. Springer-Verlag (2025). https://doi.org/10.1007/978-3-031-86185-7. Withdrawn after publication

27. Vaswani, A., et al.: Attention is all you need. In: Guyon, I., von Luxburg, U., Bengio, S., Wallach, H.M., Fergus, R., Vishwanathan, S.V.N., Garnett, R. (eds.) Advances in Neural Information Processing Systems 30 (NeurIPS), pp. 5998–6008 (2017). https://proceedings.neurips.cc/paper/2017/hash/3f5ee243547dee91fbd053c1c4a845aa-Abstract.html

28. Vreeken, J.: DeepKrimp – It's up for grabs! Personal communication (2025)

29. Vreeken, J., van Leeuwen, M., Siebes, A.: KRIMP: mining itemsets that compress. Data Min. Knowl. Disc. **23**(1), 169–214 (2011). https://doi.org/10.1007/s10618-010-0202-x

30. Zhou, L., Zhang, R., Dai, X., Hershcovich, D., Li, H.: Large language models penetration in scholarly writing and peer review (2025). https://doi.org/10.48550/ARXIV.2502.11193

Author Index

© The Editor(s) (if applicable) and The Author(s), under exclusive license
to Springer Nature Switzerland AG 2026
M. van Leeuwen and J. Vreeken (Eds.): Arno Siebes Festschrift, LNCS 16067, p. 227, 2026.
https://doi.org/10.1007/978-3-032-03028-3

The manufacturer's authorised representative in the EU is Springer
Nature Customer Service Centre GmbH, Europaplatz 3, 69115 Heidelberg,
Germany. If you have any concerns regarding our products, please
contact ProductSafety@springernature.com

Printed and bound by CPI Group (UK) Ltd, Croydon, CR0 4YY
28/04/2026
02098530-0001